Geospatial Data Science Techniques and Applications

Geospatial Data Science Techniques and Applications

Edited by
Hassan A. Karimi and Bobak Karimi

CRC Press
Taylor & Francis Group
Boca Raton   London   New York

CRC Press is an imprint of the
Taylor & Francis Group, an **informa** business

CRC Press
Taylor & Francis Group
6000 Broken Sound Parkway NW, Suite 300
Boca Raton, FL 33487-2742

Printed on acid-free paper

International Standard Book Number-13: 978-1-138-62644-7 (Hardback)

Visit the Taylor & Francis Web site at
http://www.taylorandfrancis.com

and the CRC Press Web site at
http://www.crcpress.com

Contents

Preface

For computers to be able to fully automate tasks or augment humans' decision-making capabilities, they must obtain, understand, and compute all relevant data. In our "digital" world, data come from different sources in both structured and unstructured forms, are available in different formats, and can be multi-dimensional, among other possible variations. At present, many systems and applications are of benefit to humans only when they can efficiently and effectively understand the data made available to them, differentiate between the good (useful) data and the bad (useless) data, and analyze the good data to achieve the objective in a timely manner. This latter item is very important in that "time" is a major factor in making many decisions. Thus, not only must the results of the analytics be reliable, but also their outcomes must be achieved in a timely manner to be meaningful to the decision-making effort. Producing "reliable results" and making "in-time" decisions are challenges related to Big Data as well, hence the connection between Data Science and Big Data. Self-driving cars are an excellent example: new data of various characteristics are continually provided to the decision-making component of self-driving cars and data must be understood and analyzed rapidly to make appropriate, potentially life-saving decisions. In the case of unreliable results or delayed outcomes, the data are no longer relevant or useful, leading to dangerous situations.

A simple definition of Data Science is the automatic process of using raw data from a field/discipline, and then analyzing and processing those data to produce new, meaningful, and useful information. In other words, Data Science is the automation of turning data into useful information and turning that information into knowledge. Important keywords commonly used in the context of Data Science are automation, data analytics, knowledge discovery, and prediction, among others. While many theories and techniques of Data Science have been actively researched and developed for some time now, the field of study has become quite popular recently. This is primarily due to advances in computer and information science theories and technologies as well as due to the increase in challenges of dealing with data sets that are continually growing in complexity. Data sets can be very large, of various data types, and with rapidly changing content (common characteristics of Big Data). It is worth mentioning that while Data Science has been widely discussed in the context of Big Data recently, there are several other data challenges for which Data Science techniques are needed and useful.

In short, Data Science is about computational efforts to understand and analyze data in order to make decisions, predict outcomes, or identify phenomena. While most current systems and applications are "data rich," without appropriate techniques to understand and analyze the data, there is a gap in how they can effectively assist humans in detecting and recognizing patterns and making predictions, among other important things. There are a variety of techniques that scientists may apply to the data of interest and there are often challenges in applying these techniques to those data that have complex and unique characteristics such as geospatial data.

As is clear from this broad definition, Data Science theories and techniques are needed to address complex real-world problems in a wide variety of fields/disciplines. Fields/disciplines that involve geospatial data are of particular interest due to their unique need to deal with geospatial data and analysis. To that end, this book, titled *Geospatial Data Science: Techniques and Applications*, is focused on the theories and techniques of Data Science that will benefit professionals, researchers, developers, scientists, engineers, and students interested in learning data techniques and skills to specifically address problems involving geospatial data. The scope of geospatial problems to which Data Science theories and techniques can be applied covers such diverse disciplines as environmental science and engineering, transportation planning and engineering, urban planning, social network analysis, geology, and geography, to name just a few.

This book contains 10 chapters focused on describing those Data Science theories and techniques that are particularly applicable to solving geospatial problems. The goal is to have a collection of Geospatial Data Science techniques to help the reader master new skills and understand the types of modern Geospatial Data Science theories and techniques needed to address geospatial problems. Each chapter incorporates case studies or real-world examples of applications involving geospatial data, providing a more comprehensive examination of this complex topic.

Chapter 1 defines, from a transdisciplinary perspective, the field of Geospatial Data Science. This perspective spans the three closely related disciplines of statistics, mathematics, and computer science. The chapter argues that the theories and techniques needed for data analysis and processing connect these disciplines and the result is the reduced redundant work for data scientists.

Chapter 2 explains in detail geocoding, which is a fundamental and initial operation in many geospatial projects. The chapter discusses the basics of geocoding along with the various challenges associated with the logic of geocoding. Uncertainties related to geocoding and Web-based geocoding services and geomasking are explored.

Chapter 3 discusses a deep learning technique applied to satellite images for pattern recognition, emphasizing the need for appropriate and large data sets to include labeled training samples. The chapter proposes the use of large data sets of volunteered geographic information (VGI), which are currently available through various services such as OpenStreetMap, in order to master deep learning with satellite images.

Chapter 4 discusses visual analytical approaches to analyze the movement of massive floating car data. The chapter first outlines the state of the art in floating car visual analysis at two abstract levels: point-based and trajectory-based. It then discusses several visualization methods to explore multivariate points and trajectories in interactive visual environments.

Chapter 5 discusses homology as a technique to predict patterns in geospatial data. The main feature of the technique is that it takes into account the topological properties of data in order to recognize patterns. The technique is applied to a geological data set to recognize patterns of similarity among geologic structures at tectonic boundaries.

Chapter 6 is focused on LiDAR technology and the type of data it produces. The chapter then presents a specific marine application to which LiDAR data have been applied. The LiDAR data are used in the application as the source data to develop DTM and DEM data, which are then used to quantify the effects of rises in sea levels.

Chapter 7 explains spatial-temporal techniques that are suitable for analysis of point patterns. Of the existing techniques, the chapter discusses the implementation of the spatial-temporal Ripley's K function at various scales to estimate the spatial-temporal signature of dengue fever in Columbia.

Chapter 8 focuses on geospatial data analysis of transportation demand, which depends on collecting and analyzing geospatial data, georeferenced socio-demographic data, economic data, and environmental data. The chapter reviews state-of-the art traffic data collection and analysis, as well as transportation modeling and simulation techniques.

Chapter 9 discusses utilizing data analytics to study human dynamics in the space-time context. The data analytics are applied to a water bond effort in California as the case study, measuring people's votes on the issue through tweets to explore the space-time dynamics of social media and topical distribution.

Chapter 10 presents the concept of geospatial stream processing from the user's perspective. It then discusses a framework for efficient real-time analysis of big geospatial streams, based on distributed processing of large clusters of data and a declarative, SQL-based approach.

We hope that this book will help readers, whether familiar with or new to Data Science, interested to learn about Geospatial Data Science theories and techniques with example applications involving geospatial data.

Hassan A. Karimi
Bobak Karimi

Editors

Dr. Hassan A. Karimi is a professor and the director of the Geoinformatics Laboratory in the School of Computing and Information at the University of Pittsburgh. He holds a PhD in geomatics engineering (University of Calgary), an MS in computer science (University of Calgary), and a BS in computer science (University of New Brunswick). Professor Karimi's research interests include computational geometry, machine learning, geospatial big data, distributed/parallel computing, mobile computing, navigation, and location-based services. In his research, Professor Karimi seeks to exploit the rich interdependence between the theory and practice. His research in geoinformatics has resulted in numerous publications in peer-reviewed journals and conference proceedings, as well as in many workshops and presentations at national and international forums. Professor Karimi has published the following books: *Indoor Wayfinding and Navigation* (sole editor), published by Taylor & Francis (2015); *Big Data: Techniques and Technologies in Geoinformatics* (sole editor), published by Taylor & Francis (2014); *Advanced Location-Based Technologies and Services* (sole editor), published by Taylor & Francis (2013); *Universal Navigation on Smartphones* (sole author), published by Springer (2011); *CAD and GIS Integration* (lead editor), published by Taylor & Francis (2010); *Handbook of Research on Geoinformatics* (sole editor), published by IGI (2009); and *Telegeoinformatics: Location-Based Computing and Services* (lead editor), published by Taylor & Francis (2004).

Dr. Bobak Karimi is an assistant professor of geology in the Department of Environmental Engineering and Earth Sciences at Wilkes University. He earned a BS and PhD in geology from the University of Pittsburgh. Dr. Karimi's research interests include remote sensing and GIS techniques applied to structural geology and tectonics, numerical modeling to simulate known stress/strain tests, validation of remote and modeling work by field analysis, and geophysical methods applied to increase breadth of knowledge for regional tectonic assessments. His research findings have appeared in a number of publications ranging on topics from tectonics, to geohazards, energy resources, and carbon capture and sequestration (CCS). Dr. Karimi has presented his research at many university colloquia, national, and regional conferences. His more recent research is focused on unraveling the more complex details of Appalachian tectonics and orocline development.

Contributors

Reem Y. Ali
Department of Computer Science
University of Minnesota
Twin Cities, Minnesota

Kathryn A. Bryant
Department of Mathematics
Colorado College
Colorado Springs, Colorado

Irene Casas
Department of Social Sciences
Louisiana Tech University
Ruston, Louisiana

Jiaoyan Chen
GIScience Research Group
Department of Geography
Heidelberg University
Heidelberg, Germany

Rahul Deb Das
Department of Infrastructure
    Engineering
The University of Melbourne
Melbourne, Australia

and

Department of Geography
University of Zurich
Zurich, Switzerland

Eric Delmelle
Center for Applied GIScience
and
Department of Geography and
    Earth Sciences
University of North Carolina
Charlotte, North Carolina

Linfang Ding
Applied Geoinformatics
University of Augsburg
Augsburg, Germany

and

Chair of Cartography
Technical University of Munich
Munich, Germany

Emre Eftelioglu
Department of Computer Science
University of Minnesota
Twin Cities, Minnesota

Zdravko Galic
Department of Electrical
    Engineering and Computing
University of Zagreb
Zagreb, Croatia

Alexander Hohl
Center for Applied GIScience
and
Department of Geography and
    Earth Sciences
University of North Carolina at
    Charlotte
Charlotte, North Carolina

Bobak Karimi
Department of Environmental
    Engineering and Earth Sciences
Wilkes University
Wilkes-Barre, Pennsylvania

Jukka M. Krisp
Applied Geoinformatics
University of Augsburg
Augsburg, Germany

Ajoy Kumar
Department of Earth Sciences
Millersville University
Millersville, Pennsylvania

Yu Lan
Center for Applied GIScience
and
Department of Geography and
    Earth Sciences
University of North Carolina
Charlotte, North Carolina

Shengwen Li
School of Information Engineering
China University of Geosciences
Wuhan, China

Yan Li
Department of Computer Science
University of Minnesota
Twin Cities, Minnesota

Liqiu Meng
Chair of Cartography
Technical University of Munich
Munich, Germany

Nathan Murry
Department of Earth Sciences
Millersville University
Millersville, Pennsylvania

Zahra Navidi
Department of Infrastructure
    Engineering
The University of Melbourne
Melbourne, Australia

Claudio Owusu
Center for Applied GIScience
and
Department of Geography and
    Earth Sciences
University of North Carolina
Charlotte, North Carolina

Adiyana Sharag-Eldin
Department of Geography
Kent State University
Kent, Ohio

Shashi Shekhar
Department of Computer Science
University of Minnesota
Twin Cities, Minnesota

Brian Spitzberg
School of Communication
San Diego State University
San Diego, California

Wenwu Tang
Center for Applied GIScience
and
Department of Geography and
    Earth Sciences
University of North Carolina at
    Charlotte
Charlotte, North Carolina

Xun Tang
Department of Computer Science
University of Minnesota
Twin Cities, Minnesota

Ming-Hsiang Tsou
Department of Geography
San Diego State University
San Diego, California

Stephan Winter
Department of Infrastructure
    Engineering
The University of Melbourne
Melbourne, Australia

Yiqun Xie
Department of Computer Science
University of Minnesota
Twin Cities, Minnesota

Xinyue Ye
Department of Geography
Kent State University
Kent, Ohio

Minrui Zheng
Center for Applied GIScience
and
Department of Geography and
     Earth Sciences
University of North Carolina
Charlotte, North Carolina

Alexander Zipf
GIScience Research Group
Department of Geography
Heidelberg University
Heidelberg, Germany

1

Geospatial Data Science: A Transdisciplinary Approach

**Emre Eftelioglu, Reem Y. Ali, Xun Tang,
Yiqun Xie, Yan Li, and Shashi Shekhar**

CONTENTS

1.1 Introduction

This chapter provides a transdisciplinary scientific perspective for the geospatial data science which promises to create new frontiers for the geospatial problems which were previously studied with a trial and error approach.

A well-known example from the past illustrates how rigorous scientific methods may change a field. Alchemy, the medieval forerunner of chemistry, once aimed to transform matter into gold (Newman and Principe 1998). Alchemists worked tirelessly for years trying to combine different matter and observe their effects. This trial and error process was successful for finding new alloys (e.g., brass, bronze, etc.) but not for creating another metal, that is, gold. Later, the science of chemistry showed the chemical reactions and their effects on elements, and successfully proved that an element cannot be created by simply melting and combining other elements.

We see similar unrewarded efforts (Legendre et al. 2004; Mazzocchi 2015) in the current trial and error approach to geospatial data science. We believe that research in the field needs to be conducted more systematically using methods scientifically appropriate for the data at hand.

This chapter investigates geospatial data science from a transdisciplinary perspective to provide such a systematic approach with the collaboration of scientific disciplines, namely, mathematics, statistics, and computer science.

1.1.1 Motivation

Over the past decade, there has been a significant growth of cheap raw geospatial data in the form of GPS trajectories, activity/event locations, temporally detailed road networks, satellite imagery, etc. (H. J. Miller and Han 2009; Shekhar et al. 2011). These data, which are often collected around the clock from location-aware applications, sensor technologies, etc., represent an unprecedented opportunity to study our economic, social, and natural systems and their interactions.

Consequently, there has also been rapid growth in geospatial data science applications. Often, geospatial information retrieval tools have been used as a type of "black box," where different approaches are tried to find the best solution with little or no consideration of the actual phenomena being investigated. Such approaches can have unintended economic and social consequences. An example from computer science was Google's "Flu Trends" service, begun in 2008, which claimed to forecast the flu based on people's searches. The idea was that when people have flu, they search for flu-related information (e.g., remedies, symptoms). Google claimed to be able to track flu trends earlier than the Centers for Disease Control. However, in 2013, the approach failed to identify the flu season, missing the peak time by a large margin (e.g., 140%) (Butler 2013; Lazer et al. 2014; Drineas and Huo 2016).

This failure is but one example of how the availability of a computational tool does not mean that the tool is suitable for every problem. A recent New York Times article discussed similar issues in big data analysis from the statistics perspective, concluding, "[Statistics is] an important resource for anyone analyzing data, not a silver bullet." (Marcus and Davis 2014).

Similarly, geospatial data science applications need a strong foundation to understand scientific issues (e.g., generalizability, reproducibility, computability, and prediction limits—error bounds), which often makes it difficult for users to develop reliable and trustworthy models and tools. Moreover, we need a transdisciplinary scientific approach that considers not only one scientific domain but multiple scientific domains for discovering and extracting interesting patterns in them to understand past and present phenomena and provide dynamic and actionable insights for all sectors of society (Karimi 2014).

1.1.2 Problem Definition

The term geospatial data science implies the process of gaining information from geospatial data using a systematic scientific approach that is organized in the form of testable scientific explanations (e.g., proofs and theories, simulations, experiments, etc.). A good example is USGS and NOAA's analysis of geospatial and spatiotemporal datasets, for example, satellite imagery, atmospheric data sensors, weather models, and so on, to provide actionable hurricane forecasts using statistics, machine learning (computer science), and mathematical models (Graumann et al. 2005; "National Hurricane Center" 2017).

The most important aspect of a scientific process is objectivity (Daston and Galison 2007), meaning the results should not be affected by people's perspectives, interests, or biases. To achieve objectivity, scientific results should be reproducible (Drummond 2009; Peng 2011). In other words, using the claims in a scientific study, the results should be consistent and thus give the same results every time.

Although they vary by domain (Gauch 2003), for geospatial data science we provide the following steps (Figure 1.1), which can provide objectivity and reproducibility.

The first step is the selection of a phenomenon to explain scientifically. In other words, we decide which problem we want to explain. Next, sufficient data about the phenomenon are collected to generate a hypothesis. The important aspect of this step is that hypothesis generation should be objective and not biased by scientists' perspective or interests. Experiments and simulations are then done to test the hypothesis. If the hypothesis survives these tests, then a theory can be generated. Note that in some domains, theories can be validated by mathematical proofs, and then confirmed by

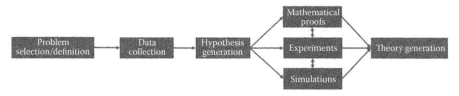

FIGURE 1.1
Steps of geospatial data science.

experiments and simulations. Thus, scientific methods differ slightly from one scientific domain to another.

This scientific process will also draw boundaries of predictability just as chemistry drew boundaries for creating matter (i.e., gold). Depending on the data in hand, non-stationarity in time may impact the success of predictability. Thus, past events may not always help predict the future. Similarly, black swan events, where the occurrence of a current event deviates from what is expected, may escape the notice of individual disciplines (Taleb 2007). The proposed transdisciplinary approach encourages us to investigate such events for better understanding the cause and predictability of black swan events with a scientific approach.

1.1.3 Challenges

Geospatial data science poses several significant challenges to both current data scientific approaches as well as individual scientific disciplines.

First, the increasing size, variety, and the update rate of geospatial data exceed the capacity of commonly used data science approaches to learn, manage, and process them with reasonable effort (Evans et al. 2014; Shekhar, Feiner, and Aref 2015). For example, vehicle trajectory datasets that are openly published on the Planet GPX web site include trillions of GPS points, each of which carries longitude, latitude, and time information ("Planet.gpx— OpenStreetMap Wiki" 2017).

Second, geospatial data often violate fundamental assumptions of individual traditional scientific disciplines. For example, in statistics, the independent and identically distributed (i.i.d.) assumption of random variables, and the stationarity assumption (whereby the mean, variance, and autocorrelation are assumed to be stationary) do not hold for geospatial data (Shekhar et al. 2015). Similarly, in mathematics, regions with indeterminate boundaries may not be represented with traditional topology and geometry, although in a geographical space indeterminate boundaries are needed since neighborhoods or urban areas often do not have determinate (strict) boundaries (Clementini and Di Felice 1996; Cohn and Gotts 1996). Also, graphs in mathematics cannot be used to represent spatial networks (e.g., road networks, rivers, etc.) since these networks have location information as well as node specific constraints (e.g., turns, traffic lights, etc.) (Barthelemy 2011). In addition, computer science often deals with one-dimensional data while geospatial data often have two, three, or more dimensions. A simple example is "sorting." In computer science, sorting may be done in one-dimensional vectors. However, there is no simple notion of sorting multidimensional geospatial data (Samet 2015).

A third challenge is that, owing to imperfect data collection devices, geospatial datasets often include missing or erroneous data (Ehlschlaeger and Goodchild 1994). To make things more complicated, there are concerns from users about geo-privacy (Kwan, Casas, and Schmitz 2004). Thus, it is hard to provide robust approaches that are generalizable.

Finally, the siloed nature of statistics, mathematics, and computer science research leads to redundant and often incomplete work on data science problems.

1.1.4 Trade-Offs

Taking a transdisciplinary view of geospatial data science means we must deal with the well-known trade-offs within individual disciplines, as well as with the many trade-offs across disciplines.

Intra-disciplinary trade-offs: An example in statistics is the tradeoff between bias and variance, as shown in Figure 1.2. A bias error occurs when wrong assumptions are used with the training dataset. In other words, during model learning we may be overly cautious, causing our model to under-fit the data, which in turn leads to a high prediction error rate. Variance error comes from the fact that even small variances in the training data are considered for model building. Such an approach may cause overfitting as well as unnecessarily complex model building and thus poor prediction performance.

An example within the discipline of computer science is the trade-off between memory storage and response time. For example, a shortest path computation using Dijkstra's algorithm (Dijkstra 1959) iteratively traverses the nodes and edges of the graph to compute the shortest path. An alternative approach may be based on precomputing and storing the shortest paths in a database with an index on the pairs of start-node and destination. This

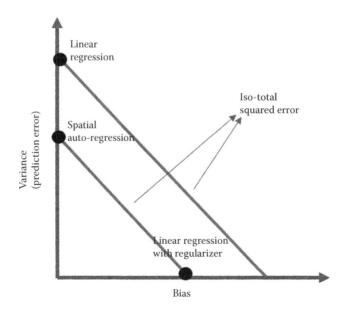

FIGURE 1.2
The trade-off between bias and variance for the statistics domain.

approach will simply use the index to retrieve the precomputed shortest path to quickly answer queries. The computation cost for shortest paths will be much lower; however, it will require much larger storage for the database of precomputed paths. The computer science literature includes many algorithms, for example, hierarchical Routing and contraction hierarchies (Geisberger et al. 2008), which explore the trade-off between storing a subset of precomputed paths and on-the-fly computation. Another computer science example from distributed systems in computer science is the CAP theorem (Brewer 2000), which states that one must choose between consistency and availability where the third concern is the partition tolerance.

Beyond the trade-offs within individual disciplines, there are new transdisciplinary trade-offs to consider across mathematics, statistics, computer science, and data-driven sciences (referred to as domain sciences).

Data-driven domain science interpretation and statistics (uncertainty quantification): Data-driven domain science interpretation and statistical uncertainty quantification have different objectives. For example, in the land cover classification problem, a decision tree (Kazar et al. 2004; Z. Jiang et al. 2012; Z. Jiang et al. 2015) or random forest (Gislason, Benediktsson, and Sveinsson 2006) approach may be used to classify remote sensing imagery to land cover type (e.g., wetland, dryland, forest, urban, rural, etc.) since the resulting models (i.e., decision trees or random forests) are relatively easy for domain scientists to interpret. However, neither the decision tree nor random forest approaches quantify uncertainty or provide a statistical confidence level for predicted land-cover classes. The alternative method is using statistical approaches such as Bayesian classifiers (Giacinto, Roli, and Bruzzone 2000). These may provide uncertainty quantification and statistical confidence but the results are not as easy to interpret due to their numerical nature. Thus, there is a need for approaches that will provide uncertainty quantification as well as ease of domain interpretation.

Computer science and statistics: Computational approaches such as data mining and machine learning tools often provide computational scalability but they may not quantify uncertainty as depicted in Figure 1.2. For example, the K-means algorithm (Hartigan and Wong 1979) for clustering is computationally efficient as it converges quickly to a local minimum on the error surface. However, it does not quantify statistical confidence in the discovered clusters. For example, it cannot determine whether the clusters discovered by K-means are better than those achieved by a random partitioning of the data set. In addition, it does not provide guarantees on the solution quality. For example, it does not tell us how the quality of a local minimum recommended by the K-means procedure compares with the quality of a global minimum on the error surface. On the other hand, the expectation maximization (EM) approaches (Dempster, Laird, and Rubin 1977) may iteratively converge to a global optimum solution; however, they seldom provide guarantees on computational cost. They cannot answer questions such as, "Is it guaranteed to terminate in a reasonable time (or will it run for an infinite time)? What is the computational complexity of the EM algorithm?" In addition, statistical approaches which aim to provide probability distributions as well as evaluate the results with statistical significance levels often require hypothesis testing (Johansen 1991), which increases the computational cost. Therefore, new research is required to provide computational scalability and statistical uncertainty quantification at the same time (Figure 1.3).

Mathematics and statistics: A pure mathematical optimization approach to estimate parameters of a statistical (or machine learning) model may lead to overfitting (Babyak 2004), which may cause the model to perform poorly on generalization for prediction on unseen datasets. Moreover, it may cause many statistical models (e.g., regression and decision trees) to become excessively complex and hard to interpret. For example, in a regression, given any set of data points, it is possible to find a polynomial function that exactly passes through each point. This may cause overfitting and reduce the prediction power of the model, since the dataset may have noisy points that bias the results. In summary, there is a need for tools that preserve statistical interpretation and mathematical completeness as well as prevent statistical models from becoming overly complex.

Mathematics, computer science, and statistics: Mathematics and statistics often have conflicting objectives. Basically, statistical inferences often involve quantifying the uncertainties with confidence intervals and statistical significance values. On the other hand, mathematics often deals with results'

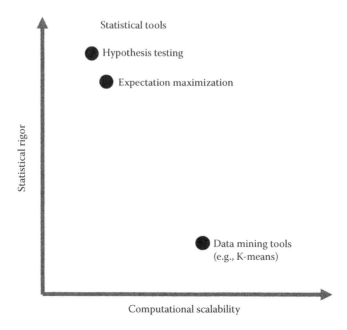

FIGURE 1.3
The trade-off between computational scalability and statistical rigor.

completeness, optimality, and so on. Many statistical methods do not guarantee mathematical properties, for example, completeness and optimality. For example, consider SaTScan (Kulldorff 1997, 1999), an algorithm to find hotspots, that is, circular areas within which the density of a phenomenon (e.g., disease, crime) is much higher than the density outside. This method uses a statistical measure, for example, a likelihood ratio and p-value, to reduce chance patterns and quantify uncertainty. This software, widely used in epidemiology for hotspot detection, enumerates circles using pairs of points, where one point defines the center and the distance between the points defines the radius. However, this approach does not enumerate many other circles, such as those defined by subsets of three points. It is likely to miss circular hotspots with empty centers as it gives up mathematical completeness to reduce computational cost. There is a need for approaches that preserve mathematical completeness while providing computationally feasible and scalable solutions.

1.1.5 Background

Previous attempts to define geospatial data science (Table 1.1) often focused on pairs of disciplines, for example, statistics–computer science, mathematics–computer science, and so on. We argue that all three disciplines should be considered to provide an understanding of naturally occurring phenomena. Moreover, these disciplines should operate together so that all may benefit from conceptual advances of common interest. For example, analytics on hyperspectral remote sensing imagery, which is used by earth science applications (e.g., agronomy, geology, hydrology, etc.), applies computationally efficient and statistically robust algorithms for those high dimensional (e.g., hyperspectral) geospatial datasets (Melgani and Bruzzone 2004).

Recently, the NSF workshop on "Theoretical Foundations of Data Science" (Drineas and Huo 2016) attempted to provide a definition of "data science" that brings these three disciplines together. The workshop identified fundamental areas where collaboration among computer scientists, mathematicians, and statisticians is necessary to achieve significant progress. However, the focus of the workshop was not geospatial data generally but high-dimensional data, and most of the discussion centered on very specific topic areas, that is, computation-statistics tradeoff, randomized numerical linear algebra,

TABLE 1.1

Overview of Related Work

	High-Dimensional Data	Spatial Data
Siloed/ multidisciplinary	Statistics, mathematics, and computer science	Spatial statistics, Spatial data mining, and machine learning
Transdisciplinary	Theoretical foundations of data science workshop	Proposed approach

signal processing/harmonic analysis on graphs, nonconvex statistical optimization, combining physical and statistical models, mixed type and multimodality data, applied representation theory and noncommutative harmonic analysis, topological data analysis (TDA) and homological algebra, security, privacy, and algorithmic fairness (Drineas and Huo 2016).

Another recent development was the NSF Workshop on Geospatial Data Science in the Era of Big Data and CyberGIS ("NSF Workshop on Geospatial Data Science" 2017). Its focus was high-performance computing and the computational aspects of geospatial data science. Topics included geospatial big data capabilities (e.g., LiDAR, remote sensing, and location-based social media) for novel applications (e.g., urban sustainability), cloud computing, and tools for scalable geospatial data analytics. One of the goals was to formulate a core set of questions and problems of geospatial data science around these themes. The workshop addressed the geospatial data science problem from a high-performance computing perspective but did not address the broader set of questions that led us to our attempt here to define geospatial data science.

1.1.6 Contributions and the Scope and Outline of This Chapter

This chapter takes a wide-lens perspective on geospatial data science. We believe that geospatial data science is a transdisciplinary field comprising statistics, mathematics, and computer science, and that it should be formally considered the foundation of geospatial science. The aim is both to reduce redundant work across disciplines as well as to define the scientific boundaries of geospatial data science so it is no longer seen as "a black box" solution to every possible geospatial problem. In addition, we aim to lay out some of the challenges that arise from the geospatial nature of the data. Hence, in the following sections, we investigate individual disciplines, their objectives as well as the challenges they face to investigate the transdisciplinary definition of geospatial data science.

Scope and outline: In this chapter, we present geospatial data science as a transdisciplinary scientific process. The proposed approach provides a discipline-of-disciplines perspective toward reducing redundant work and providing a more robust way to create information from raw geospatial data. In addition, our approach aims to identify the limits of geospatial data science predictability.

To emphasize the transdisciplinary perspective of geospatial data science, in the following sections we provide examples from each discipline, namely, statistics, mathematics, and computer science, that are cross-cutting with geospatial data science. As summarized in Figure 1.4, for example, the study of indeterminate regions is both a mathematics and a spatial statistics problem. Similarly, randomized algorithms can be considered not only as a problem in computer science but also one that uses fundamental ideas from spatial statistics. Finally, representative problem examples that all three disciplines tackle are explained in more detail.

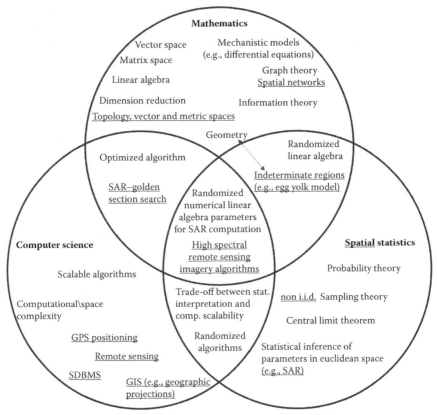

FIGURE 1.4
Comparison of disciplines via examples (one theory/approach doesn't fit all).

1.2 Statistics

1.2.1 Traditional Statistics

Statistics studies data in the context of uncertainty, and it serves as an important foundation of many data science tasks such as pattern recognition, prediction, and classification. Given the observations collected from a part of a population, statistics reduces uncertainty by making inferences on the entire population. It differs from probability theory in that probability theory works with known probabilistic distributions to estimate the probability of future observations while statistics starts with a collection of past observations and estimates the unknown parameters of a probabilistic distribution to make inferences.

In statistics, data collection is performed based on sampling theory (Härdle, Klinke, and Röonz 2015), which provides a scientific framework to decide the population of concern, sampling approach (e.g., random sampling), sampling size, and so on. The collected observations are then used to estimate parameter values of a target distribution model (e.g., Gaussian distribution). The estimation can be performed using either a "frequentist" or a Bayesian approach (Hong et al. 2013). A frequentist approach analyzes data as an integrated whole. It assumes each parameter has a fixed value that does not change over time and that can be accurately estimated as the number of observations increases to infinity. However, in real-world scenarios, the number of observations is limited and there is always an uncertainty associated with the analysis, given the incomplete data. In order to express this uncertainty, a frequentist approach uses a confidence interval (Curran-Everett 2009) to claim a minimum expected probability (e.g., 95%) that the estimated parameters are true.

By contrast, a Bayesian approach assumes that each parameter comes from a prior distribution. It considers data as a sequence of observations and continues to update the estimation of parameters as new observations are available. Unlike a frequentist approach, a Bayesian approach captures the change or evolution of parameters over a sequence (e.g., time) of observations, and thus can further reduce the uncertainty in an inference. However, a Bayesian approach requires an appropriate prior distribution as input; otherwise, it cannot give correct inferences.

1.2.2 Traditional Statistics versus Spatial Statistics

One of the most common assumptions in traditional statistics is that observations are identically and independently distributed (i.i.d.) (L. Cam and Yang 2000). The i.i.d. assumption is an important foundation of many data science methods. For example, in machine learning, maximum likelihood estimation (Pan and Fang 2002) is used to estimate the parameter values of a given model, and the expressions of likelihood functions are often obtained based on this i.i.d. assumption (e.g., Naïve Bayes classifier, expectation–maximization). In fact, many classic statistics theorems come from the i.i.d. assumption, such as the well-known central limit theorem (Rice 1995), which states that the mean of a set of samples is approximately equal to the mean of an entire population, given a sufficiently large sample size.

Although it offers great convenience in traditional statistics, the i.i.d. assumption is often violated in the geospatial domain. As the first law of geography states: "Everything is related to everything else, but nearby things are more related than distant things" (Tobler 1970). This fundamental observation on geospatial data breaks the i.i.d. assumption of nonspatial data in traditional statistics. Spatial statistics deals with the phenomenon of spatial autocorrelation through careful modeling of spatial relationships among data samples. The following discusses two motivating examples of spatial statistics.

Example 1.1: Pearson Correlation on Geospatial Data

Figure 1.5a shows a distribution of three types of diseases, abbreviated as TF, VSD, and ALL. Each instance of each disease has a unique ID as marked in Figure 1.5a. From the distribution, we can see each ALL instance has a nearby TF instance and VSD instance. For example, ALL1 is adjacent to TF1 and VSD1. To measure the spatial correlation among the three types of diseases, we need some parameters to express the spatial distribution. Figure 1.5b shows a boundary fragmenting the study area. For each type of disease, we can consider each fragment as a property of its spatial distribution, and each property value as the count of instances of this disease within the fragment. Suppose the fragments are concatenated into a vector following column-wise order (top-left → bottom-left → top-right → bottom-right). Thus, the vector of properties for ALL is [0, 0, 1, 1], TF is [1, 2, 0, 0], and VSD is [0, 0, 1, 1]. With this spatial modeling based on boundary fragmentation, the Pearson correlation ratio is –0.91 between TF and ALL, and 1 between VSD and ALL. This negative correlation between TF and ALL contradicts our observation since their spatial adjacency is broken by the boundary between fragments (Figure 1.5b). By contrast, the correlation between VSD and ALL is positive because the spatial adjacency between VSD and ALL instances is preserved by the arbitrary partitioning. These mutually contradictory correlations reveal the uncertainty of results when traditional statistics is trivially applied to the geospatial domain.

Example 1.2: Agronomic Field Experiment Design

Field experiments are used by agricultural scientists to evaluate the performance and properties of crops under different conditions (e.g., water and fertilizer) (Legendre et al. 2004; Van Es et al. 2007). Traditional experiment designs assume that observations are independent and that the expected value stays the same at different spatial locations. However, in field experiments, these assumptions are often violated since closer plants exhibit more similar properties, and soil properties vary at different locations, which lead to nonstationary expectations (Legendre et al. 2004). To address this problem, blocks are used in field experiment

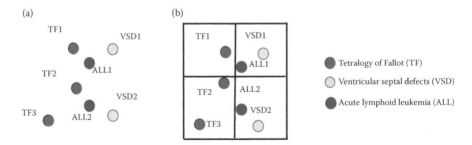

FIGURE 1.5
Distribution of disease. (a) A map of 3 types of diseases and (b) Boundary fragmenting the study area.

design to reduce the effect of spatial autocorrelation and heterogeneity. A block is a large spatial unit containing a set of plots. With a properly chosen block size, the spatial-related properties (e.g., soil type) can be assumed to be uniform within a block. Distances are added between blocks so that the spatial autocorrelation between blocks is reduced. The choice of block size and distance between blocks are critical parameters to reduce the errors caused by spatial effects. In practice, they can be determined by spatial statistical analysis.

1.2.3 Spatial Statistics

Geostatistics: Geostatistics (Chiles and Delfiner 2012) is concerned with point-reference data, which contains a set of points with fixed locations and a set of attribute values. The goal of geostatistics is to model the distribution of the attribute values and make predictions on uncovered locations. Point-reference data have several inherent properties: (1) isotropy/anisotropy; (2) second-order stationarity; and (3) continuity. In the context of isotropy, uniformity is assumed in all directions, while under anisotropy, some statistical properties may vary by direction. Second-order stationarity is a weaker form of strong stationarity, so it is also referred to as weak stationarity. Instead of assuming a strong stationarity with invariant density of distribution, second-order stationarity assumes only invariant moments (e.g., mean and variance) across a spatial domain but covariance between locations depends on the distance. The continuity property indicates the existence of spatial dependence on the data. The degree of dependence can be quantitatively measured with input distance and direction using a variogram or semivariogram. If we further assume isotropy, then the variogram simplifies to a function of distance only. With the base assumptions on point-reference data, the distribution of attribute values can be effectively modeled. A set of statistical tools is provided by geostatistics and one of the most popularly used methods is Kriging (Williams 1998). Kriging is a statistical model of interpolation that predicts attribute values at unsampled locations (e.g., water quality estimation based on observations from a set of monitoring sites). Co-Kriging (Stein and Corsten 1991) provides a multivariate extension of ordinary Kriging. For a set of highly correlated attributes, Co-Kriging can improve the prediction quality on a poorly sampled attribute using well-sampled ones. Besides spatial autocorrelation, spatial heterogeneity also needs careful consideration in many applications (e.g., different types of underlying landscape). Special models, such as GWR (geographically weighted regression) and spline, are available in geostatistics to reflect the changes in statistical properties, given the presence of spatial heterogeneity. These models deploy a local view on the data and assign higher weights to neighboring points to reduce the effect of heterogeneity.

Spatial point process: Unlike geostatistics, a spatial point process is not concerned with attribute values but with the locations of points (Møller and Waagepetersen 2007), specifically their distribution. Locations of a set

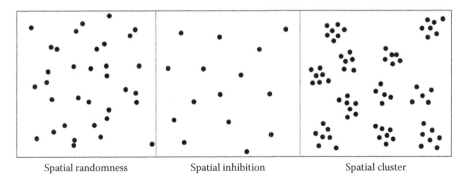

| Spatial randomness | Spatial inhibition | Spatial cluster |

FIGURE 1.6
Forms of spatial point process.

of points can be generated based on different statistical assumptions (e.g., random and clustered). The most common model assumed for a spatial point process is a homogeneous Poisson distribution, also known as complete spatial randomness (CSR). In CSR, the total number of points follows a Poisson distribution and each point is identically and independently distributed in a predefined spatial domain. A variant of CSR is a binomial point process, in which the only difference is a fixed total number of points. In many application domains, CSR or binomial point process is not an appropriate assumption since points may have spatial autocorrelation or inhibition characteristics. In such cases, other specialized models should be applied to better approximate the exact distribution as shown in Figure 1.6. For spatial inhibition, a Poisson hardcore process is widely used to generate a distribution that enforces mutual repulsion among points. For spatial autocorrelation, a Matern cluster process can be chosen to reflect the clustering characteristics. Similar cluster processes include the Poisson cluster process, Cox cluster process, Neyman-Scott process, and so on. One of the most well-known applications of a spatial point process is spatial scan statistics (Kulldorff 1997; Agarwal et al., 2006; Neill and Heinz 2009; E. Eftelioglu, Tang, and Shekhar 2015; Tang, Eftelioglu, and Shekhar 2015; E. Eftelioglu, et al. 2016b,c) in hotspot detection. In spatial scan statistics, chance hotspots are removed through a statistical significance test under a null hypothesis based on CSR. CSR is also used as a null hypothesis for significance testing in Ripley's K function (Dixon 2002), which estimates the overall clustering degree of a point distribution.

Lattice statistics: A lattice is a representation of a discrete space, which is a finite collection of grid cells in a spatial domain. In this case, lattice statistics concerns statistical processes in the field model. For continuous data (e.g., polygon), a W-matrix (continuity matrix) can be computed to transform the original data into a discretized representation based on their spatial adjacency or proximity. Lattice statistics provides a set of models (Cliff and Ord 1981; Getis and Ord 2010), such as Moran's I, Getis-Ord Gi*, Geary's

C, Gamma index, and LISA, to evaluate spatial autocorrelation on the field model. For example, Moran's I outputs an I-value within [−1, +1] to reflect a positive, none, or negative spatial autocorrelation in the input dataset. For value estimation and prediction, spatial autoregressive models (M. M. Wall 2004) are applied on discrete data, such as Markov random fields (MRF), the simultaneous autoregressive model (SAR), and the conditional autoregressive model (CAR). MRF models the evolution process of a phenomenon based on the assumption that the property of a spatial location is spatio-temporally determined by its neighbors with additional randomness. In the CAR model, a Markov property is implied and the state of a location is affected by its direct neighbors, but not neighbors of its neighbors. This property of CAR is called spatial memoryless. By contrast, SAR does not assume any non-transitive spatial influences and considers autocorrelation in a larger spatial domain. Therefore, CAR is a more appropriate choice for a local spatial process and SAR is a better assumption for a global spatial process. Another critical issue in lattice statistics is the impact of scale on spatial analysis. With different aggregation levels of scale, the statistical analyses may have distinct results. For example, variance of income aggregated on a neighborhood level could be much smaller than that on a county level within the same state.

Spatial network statistics: A spatial network is a graph-based model with enriched spatial information (e.g., turn and capacity). In a spatial network, events or objects are mutually accessed through a set of connected edges instead of straight lines in the Euclidean space. Statistics on spatial networks is a newly emerging area which has not been as extensively studied as statistics on Euclidean space. In recent work, some statistical models for object data, such as spatial autocorrelation, interpolation, and clustering approaches, have been extended to spatial networks. Spatial network statistics, as an extension of spatial statistics on Euclidean space, can better model processes in urbanized places where objects and events spread along network edges (e.g., roads and rivers). For example, in transportation planning, statistically significant hotspots of accidents need to be identified based on network space (Tang et al. 2017).

1.3 Mathematics

Mathematics plays a critical role in all science and technology. It is fundamental to a variety of traditional subjects such as physics, chemistry, and agriculture. In data science, mathematics provides its core value in data representation and modeling as well as the logic and proofs used to validate data science approaches. In this section, we first introduce how mathematics is applied in traditional data science with a collection of examples. Then we discuss the limitations of applying traditional mathematical models to spatial data and novel spatial models with examples.

1.3.1 Mathematics in Traditional Data Science

Data science utilizes a variety of subjects for accomplishing different data modeling and processing tasks. Many types of data can be represented using linear algebra models. Aligned two-dimensional data are typically modeled as a matrix. For example, an image channel is represented as a matrix where each element indicates the value of a pixel in the corresponding location. This representation is widely applied in precision agriculture (Campbell and Wynne 2011; Lillesand, Kiefer, and Chipman 2014; E. Eftelioglu et al. 2016a), which discovers the remote sensing data consisting of multiple image channels. In addition, a graph can be represented as a neighborhood matrix as well where each row corresponds with a node in the graph and the elements in that row indicate the connection from this node to all the other nodes. A vector is always used to model an object that has a set of feathers where each feather is quantized as an element in the vector. The operations on matrices and vectors also apply on the represented data. For example, the similarity between two feather vectors can be measured by the distance computed by the norms and the angle between them. Eigenvalue and eigenvector are used for studying the behavior of Markov chains (Gabriel and Neumann 1962; Brooks et al. 2011) which has been the core idea of many approaches such as PageRank (Page et al. 1999). Principal component analysis (PCA) (Jolliffe 2002) uses eigenvalues and eigenvectors for reducing the dimensionality of the data. Another important application of linear algebra in data science is regression (Neter et al. 1996). A linear regression can be modeled as a linear system which can possibly be solved by multiple linear algebra approaches such as Gaussian elimination and multiplying by inverse (Wilkinson and Wilkinson 1965). Many data science approaches are derived based on linear algebra. As an example, low-rank matrix approximation based on Singular Value Decomposition (SVD) (Golub and Reinsch 1970; M. E. Wall, Rechtsteiner, and Rocha 2003) is applied in data compression, classification, regression, clustering, and signal processing, and so on.

Another subject in mathematics that is widely used in data science is information theory. Entropy is a concept that originally comes from thermodynamics (Guggenheim 1985) which measures the number of microscopic configurations that a thermodynamic system has. On the basis of the essence of entropy, entropy in information theory (Ayres 1997) measures the expected value of the information contained in a message or the uncertainty of the data. Data classification approaches such as decision trees use entropy to measure the information gain (Quinlan 1986; Safavian and Landgrebe 1991; Hall and Holmes 2003) between two levels of the tree which offers a quantitative guide of how the tree should grow. For example, a good growth of the tree is expected to decrease the overall entropy.

Optimization is a highly interdisciplinary subject related to both mathematics and computer science. It is applied to many critical societal applications. For example, precision agriculture researchers need to allocate each field with a type of product to achieve the optimal environmental and economic outcome, which requires solving a multi-variable optimization problem (N. Zhang, Wang, and Wang 2002; McBratney et al. 2005). Many machine learning approaches use optimization techniques to achieve their goals such as finding the minimal value of the cost function (Govan 2006; Boyd et al. 2011). For example, gradient descent, a popular approach, finds the minimum value of a cost function by iteratively moving along the direction of the slope (L.-K. Liu and Feig 1996; Mason et al. 1999; Bottou 2010). Finding the slope requires solving differential equations (K. S. Miller and Ross 1993), which is an important subject in mathematics. Differential equations have many other applications in data science especially on spatial data, since they can be naturally differentiated into variations over space and time. For example, the Soil and Water Assessment Tool (SWAT) (Gassman et al. 2007; Douglas-Mankin, Srinivasan, and Arnold 2010) is a software that embraces a variety of environmental and agricultural models about the variations over space and time which apply differential equations.

A mathematics subject tightly related to computer science and data science is graph theory. This is because many real phenomena can be naturally modeled by a graph where the vertices represent the objects, and the edges represent the relationship between objects. For example, as a web network model (Kleinberg et al. 1999; Broder et al. 2000), each webpage can be modeled as a vertex and the links are modeled as edges outgoing from this vertex. In a social network model (Freeman 1978; Mislove et al. 2007), vertexes represent individuals, and edges represent the relationship between two individuals. There are also spatial data models based on graph theory. Traditionally, road networks are modeled such that the intersections are vertices and the roads are edges. A similar framework also applies to flight networks (Li-Ping et al. 2003) and oil pipeline networks (Brimberg et al. 2003), but the edges become the air routes and the pipelines.

Topology studies the properties that are preserved under deformations, including stretching, twisting, and bending. TDA (Zomorodian 2012) is an example of applying topology in data science whose main goal is to study the geometric characteristic of data via topology. For spatial data, they are largely used in modeling a collection of relationships between real-world spatial objects. For example, Minneapolis is inside of Minnesota state is an "inside" topological relationship.

1.3.2 Limitations of Applying Traditional Mathematical Models to Spatial Data and Novel Spatial Models via Examples

We reviewed the mathematical subjects that have been applied in data science. However, they have many non-negligible limitations when dealing with spatial data. An example comes from the metric of objects. Suppose there are two spatial objects on a two-dimensional plane, each presented by a two-dimensional coordinate; how do we order them? One straightforward way is using their distance to the origin. Another popular way is sorting by the angle between the line connecting the points and the origin and an axis (i.e., x-axis or y-axis). The point is, there is no natural metric that can order spatial points. Developing meaningful and efficient ordering metrics for spatial objects is an important and challenging research topic.

In traditional topology, spatial regions are always modeled with determinate boundaries (Randell and Cohn 1992; Cohn and Gotts 1996). It turns out that the traditional topological relationship models always rely on the boundary. For example, the relationship "inside" is determined by whether a spatial region falls completely within another region, and the relationship "touches" is determined by whether the boundaries of two spatial regions are overlapped but not their inside such as two neighbor states. However, in real-world scenarios, many spatial regions are surrounded by indeterminate boundaries. For example, it is impossible to clearly define the boundary between urban and rural areas. Research has been done to narrow the gap between real-world relationships between spatial regions and traditional topological models. One of the most popular models is the "Egg-Yolk" (Cohn and Gotts 1996) model which provides a representation of regions with indeterminate boundaries based on the framework of "RCC-theory" (Randell and Cohn 1992; Cohn, Randell, and Cui 1995). It is a logically consistent and computationally tractable model that represents a spatial region with an indeterminate boundary by pairs of regions with determinate boundaries (i.e., crisp regions).

Traditionally, spatial data have always been modeled on Euclidean space. This works well for many problems such as those related to air and ocean. However, there are many activities associated with transportation networks such as traffic and crimes. Using traditional models based on Euclidean space significantly affects the precision of the model and thereby the quality of the solution. As an illustration, Figure 1.7 shows a map of the campus of University of Minnesota. The east and west banks are connected by a bridge over the Mississippi river. The Euclidean distance between the two red dots is short, yet the network distance computed from the shortest path is much longer (Dijkstra 1959). If we want to approximate the travel time between these two dots, the error using Euclidean distance will be huge.

Models based on networks space can give a better distance approximation to some extent. In the simplest way, a transportation network can be represented as a graph, where each intersection is a vertex and each road segment is an edge associated with a value representing the travel cost of that edge. The travel cost could be assigned on the basis of various values such as road distance, travel time, or fuel consumption. However, traditional graph models have several major limitations dealing with the massive information contained in spatial networks. For example, the traditional models simply

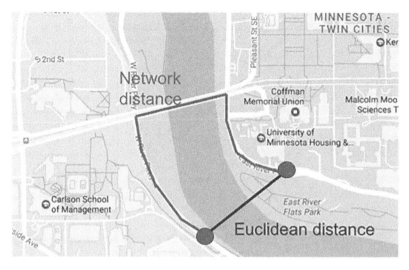

FIGURE 1.7
An illustrative example shows the difference between distances in Euclidean space and network space.

treat intersections as vertices but do not model the turns. However, according to the laws of traffic, left turns usually cost much less than right turns if driving on the left side (Lovell 2007). This difference cost could lead to serious results in real-world applications. UPS saved 10 million gallons of fuel, emits 22 thousand tons less carbon dioxide, and delivers 350 thousand more packages every year by avoiding a left turn since the year 2004 (Lovell 2007). Figure 1.8 shows an example of modeling the turns. The left figure shows a patch of map in Dinkytown, Minneapolis. The middle figure shows a traditional model describing the streets where the vertices are the intersections and the directed edges are the roads. The right figure shows an example of modeling the intersection at N_5 while keeping the turn information by a set of connects. The other approaches include using hyper-edges along with hyper graphs and annotating the graph with turn information.

In addition, in traditional graph models, each edge is associated with a static value, which is not enough for modeling dynamically changing travel costs. For example, the travel time for a highway around downtown varies a lot during rush hour and non-rush hours. A time-expanded-graph (TEG) (Köhler, Langkau, and Skutella 2002; Silver and De Weck 2007) is one approach that is capable of modeling dynamically changing weights on edges. Figure 1.9 shows an example of TEG of a graph consisting of four nodes. The left side shows the varying travel times associated with each edge in four timestamps. The right side shows the TEG modeling this graph where each column represents the set of vertices in one timestamp. Each edge connects the nodes that are reachable within a certain time. For example, edge $\langle A_1, B_3 \rangle$ indicates that if departing from Node A at timestamp 1, you will be

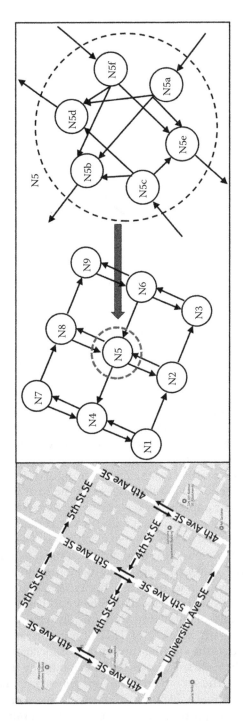

FIGURE 1.8
An example of modeling the turns of a transportation network.

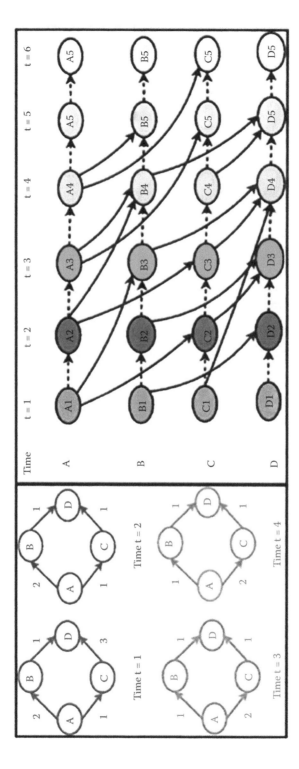

FIGURE 1.9
Example of a TEG.

arriving at Node B at timestamp 3. As can be seen, TEG is much more complex compared to the traditional static graph model, and thus leads to harder computational challenges.

Moreover, in a traditional graph, the edges are considered atomic, which cannot be further fragmented. This feather works when using graphs to model nonspatial networks such as webpage networks and social networks since the edges virtually represent the connection between objects. However, for spatial data models, the edges represent roads on which activities happen. If we treat edges as atomic, the location information of the activities will be lost. A novel model called dynamic segmentation (Dueker and Vrana 1992; Chang 2006) has been proposed to handle this limitation.

The original graph is segmented based on the locations of activities on the edges. Figure 1.10 shows an illustrative example, using traditional graph model, edge $\langle N_1, N_2 \rangle$ is atomic and the location information of activities A_1, A_2, A_3, A_4 can not be preserved due to this atomicity. In dynamic segmentation, edge $\langle N_1, N_2 \rangle$ is segmented to $\langle N_1, A_1 \rangle$, $\langle A_1, A_2 \rangle$, $\langle A_2, A_3 \rangle$, $\langle A_3, A_4 \rangle$, $\langle A_4, N_2 \rangle$, and thus the locations of the activities are kept. Using dynamic segmentation outperforms traditional models, especially when dealing with activities

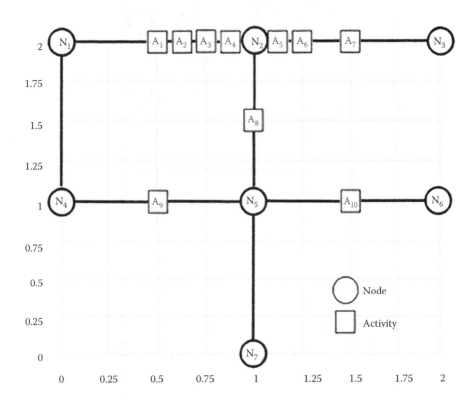

FIGURE 1.10
An illustrative example of dynamic segmentation.

located on a portion of long road segments such as a highway. For example, in linear hotspot detection, dynamic segmentation helps increase the precision of the hotspots and reveals hotspots that are missed using traditional models.

1.4 Computer Science

In this section, we start by discussing core questions and goals of computer science. We then present some of the concepts, theories, models, and technologies that computer science has contributed to the field of data science. Finally, we discuss the limitations of traditional data science with respect to spatial data and the computer science accomplishments that have attempted to address these limitations toward the realization of geospatial data science.

1.4.1 Core Questions and Goals

Computer science is both a scientific and an engineering discipline (Abrahams 1987). Hence, computer science contributions encompass both theory (e.g., studying properties of computational problems) and practice (e.g., systems design and data mining). However, the scientific aspect of computer science is different from physical disciplines and closer to mathematics where the goal is to create representation models and study their properties. Many traditional data science questions are studied within the field of computer science. Examples include: Is a given problem decidable (i.e., computable)? Is there a polynomial time algorithm to solve a given problem? What is the most efficient algorithm to perform the computations? Can the algorithm scale to large datasets? What is the tradeoff between the computational scalability of the algorithm and the statistical rigor?

1.4.2 Concepts, Theories, Models, and Technologies

Many computer science concepts are leveraged in data science. Two major concepts are the design of appropriate *data structures and algorithms*. Data structures are ways of storing data so that they can be efficiently used. Examples of common data structures include arrays, queues, linked lists, trees, and graphs. Algorithms (Cormen 2009) are well-defined computational procedures that take a value (or a set of values) as input, and produce a value (or a set of values) as output to solve a given problem (e.g., searching, sorting, and finding the shortest path between a source and a destination node in a transportation graph).

In addition, computer science theories are also leveraged in traditional data science. For instance, *computational complexity theory* (Papadimitriou

2003; "Computational Complexity Theory" 2017) focuses on classifying computational problems according to their inherent difficulty. The theory introduces mathematical models and techniques for studying computational problems and is usually used to establish proofs that, for a given problem, no algorithm can run faster than the current one.

Another major accomplishment relevant to data science is the development of *database management systems (DBMS)*, general-purpose software systems that facilitate the processes of defining, constructing, manipulating, querying, and sharing databases among users and applications (Elmasri and Navathe 2015). The most common type of DBMS is relational database management systems (RDMS), which adopt the relational data model first introduced in (Codd 1970). In this model, the database is represented as a collection of relations (i.e., tables), based on the concept of mathematical relations. Each row (i.e., tuple) typically represents information about a real-world entity or relationship, while each column represents a given attribute describing that entity. SQL is the standard query language for commercial RDBMSs and is based on relational calculus. Relational algebra is also used as the basis of query processing and optimization in RDBMS (Elmasri and Navathe 2015). Examples of popular commercial RDBMSs include IBM's DB2, Oracle, Sybase DBMS, SQL Server, Access, MySQL, and PostgreSQL.

Cloud computing platforms make possible the processing of large data volumes in an efficient manner. Existing approaches to cloud computing provide a general framework for distributed file systems (e.g., Google file [Ghemawat, Gobioff, and Leung 2003] system and HDFS [Borthakur 2007]) and processing these data sets based on replicas of data blocks (e.g., map-reduce [Dean and Ghemawat 2008], Hadoop [Borthakur 2007], and Spark ["Apache Spark™—Lightning-Fast Cluster Computing" 2017]). Figure 1.11 (left side) shows the Intel distribution for Apache Hadoop software components (Intel 2013). It also shows many components running on top of the HDFS for distributed processing (MapReduce), workflow (Oozie), scripting (Pig), machine learning (Mahout), SQL queries (Hive), and column store storage (HBase). In addition to cloud computing platforms, there are also many existing *high-performance scientific computing* cluster technologies as depicted on the right side of Figure 1.11. These computing technologies include parallel file systems (e.g., Lustre), batch schedulers (e.g., SLURM), MPI, and OpenMP for internode and intra-node parallelism, and numerical and domain specific libraries, on top of which applications are usually developed using languages such as FORTRAN and C/C++ (Reed and Dongarra 2015).

Another major area of interest in computer science is data mining. Data mining refers to the discovery and extraction of new and useful information (e.g., patterns or rules) from large amounts of data. Typically, data mining has been mainly concerned with the computational complexity of proposed discovery algorithms and less concerned with the statistical robustness of these algorithms (e.g., bias, inference confidence, etc.). Common data mining tasks include the discovery of association rules (e.g., which grocery store

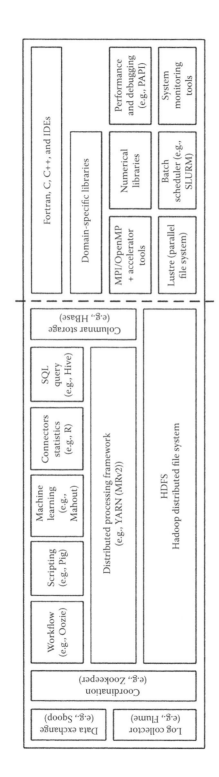

FIGURE 1.11

Intel Distribution for Apache Hadoop software components compared with the high-performance computing ecosystem. (Figure adapted from Intel. 2013. Intel distribution for Apache Hadoop software. http://www.intel.com/content/dam/www/public/us/en/documents/articles/intel-distribution-for-apache-hadoop-product-brief.pdf; Reed, D. A., and J. Dongarra. 2015. *Communications of the ACM* 58(7): 56–68.)

items are frequently bought together?). Algorithms such as Apriori (Agrawal, Srikant, and others et al. 1994) and FP-growth (Han, Pei, and Yin 2000) have been proposed for efficiently mining association patterns. Data mining tasks also include the classification problem (e.g., classifying a pixel in a picture as dryland versus wetland based on other pixel properties). Popular classification models include decision trees for which computational algorithms such as ID3 (Quinlan 1986) have been proposed.

1.4.3 Limitations of Traditional Data Science for Spatial Data and Related Computer Science Accomplishments

Now we review the limitations of traditional data science with respect to spatial data by focusing on three main areas of accomplishments, namely, spatial databases, spatial cloud-computing platforms, and spatial data mining.

Spatial databases: Applications such as precision agriculture require special database support to store, process, and query spatial data (e.g., storing and querying the polygons representing farm plots). Before the development of spatial databases, spatial queries (e.g., Which galaxy pairs are within 30 arc seconds of each other? Which houses are most likely to be flooded by global warming-induced sea-level rise?) required extensive programming and suffered from long computation times due to the mismatch between 2D spatial data and 1D data types (e.g., number) and indexes used by traditional database systems (such as B+ Tree) (Shekhar, Feiner, and Aref 2015). In addition, a naive collection of spatial data types is inadequate for multistage queries since the result of some queries (such as the union of disjoint polygons) cannot naturally be represented as a point, line, or polygon. Spatial databases (such as Oracle Spatial and PostGIS) introduced spatial data types (such as OGIS simple features ("Welcome to the OGC|OGC" 2017), operations (such as inside and distance), spatial data structures (such as Voronoi diagrams), and algorithms (such as shortest-path, nearest-neighbor, and range query) to represent and efficiently answer multistage concurrent spatial queries (Shekhar, Feiner, and Aref 2015). The reduced programming effort resulted in more compact code and quicker response times. In addition, spatial indexes have also been added. Representative indexes for point objects include Grid files, multidimensional grid files (Lee et al. 1997), Point-Quad-Trees, and Kd-trees (Samet 1990). Representative indexes for extended objects include the R-tree structures (Guttman 1984). The R-tree is a height balanced natural extension of the B+ tree for higher dimensions (Shekhar et al. 1999). Objects are represented in the R-tree by their minimum bounding rectangles (MBRs). Non-leaf nodes are composed of entries of the form (R, child-pointer), where R is the MBR of all entries contained in the child-pointer. Leaf nodes contain the MBRs of the data objects. To guarantee good space utilization and height-balance, the parent MBRs are allowed to overlap. Many variations of the R-tree structure exist whose main emphasis is on discovering new strategies to maintain the balance of the tree in case of a split and to minimize the overlap of the MBRs in order to improve the search time.

Spatial computing platforms: Support for spatial data (e.g., spatial indexes) was also needed in cloud computing platforms to improve the I/O cost of spatial queries (e.g., retrieving a set of farm polygons within a given spatial range). Representative efforts for supporting spatial data in existing cloud computing platforms include (Evans et al. 2013): (1) Spatial Hadoop (Ali and Mokbel 2017), which is a MapReduce extension to Apache Hadoop designed especially to work with spatial data by providing specialized spatial data types, spatial indexes, and spatial operations; (2) Hadoop GIS, a high-performance spatial data warehousing system over MapReduce (Aji et al. 2013); and (3) GeoSpark (Yu, Wu, and Sarwat 2015), the spatial extension for Apache Spark. Research on parallel R-tree construction on a GPU is also ongoing (Prasad et al. 2013). At the Hadoop Distributed File System (HDFS) level, Spatial Hadoop (Ali and Mokbel 2017) and Hadoop GIS (Aji et al. 2013) have added spatial indexes. At the scripting layer (e.g., Pig), Spatial Hadoop has added Open Geodata Interoperability Specification (OGIS) data types and operators. GIS on Hadoop (Pang et al. 2013) has also added OGIS data types and operators at the SQL query level (e.g., Hive). In addition to the spatial extensions of Hadoop, the GeoSpark (Yu, Wu, and Sarwat 2015) system has also extended Apache Spark with a set of Spatial Resilient Distributed Datasets (SRDDs) that can efficiently load, process, and analyze SBD. GeoSpark also introduced spatial indexes, spatial geometric operations that follow the Open Geospatial Consortium (OGC) standard, and spatial query operations for SBD.

Spatial data mining: Spatial data mining (Stolorz et al. 1995; Shekhar and Chawla 2003) is the process of discovering interesting and potentially useful patterns from spatial databases. For example, in precision agriculture, given a UAV-captured image of a farm, one may want to classify the set of pixels in the image based on the crop type (e.g., corn, soybean, etc.).

However, the complexity of spatial data and implicit spatial relationships limits the usefulness of conventional data mining techniques for extracting spatial patterns (Shekhar et al. 2011). Specific features of geographical data that preclude the use of general purpose data mining algorithms are (1) the spatial relationships among the variables, (2) the spatial structure of errors, (3) the presence of mixed distributions as opposed to commonly assumed normal distributions, (4) observations that are not independent and identically distributed (i.i.d.), (5) spatial autocorrelation among the features, and (6) nonlinear interactions in feature space. Figure 1.12 (Z. Jiang et al. 2013) illustrates an example of these limitations, namely the existence of spatial autocorrelation, by comparing the output of traditional decision trees with spatial decision trees for classifying wetland and dryland pixels in a satellite image taken in the city of Chanhassen, MN. The classification model used 12 continuous explanatory features as input, including multi-temporal spectral information (R, G, B, NIR bands) and normalized difference vegetation index

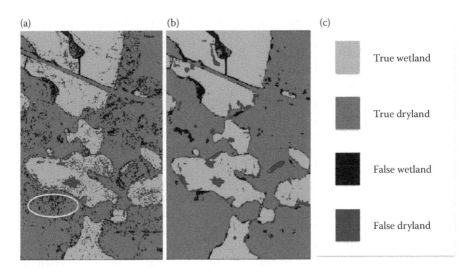

(a)   (b)   (c)

True wetland

True dryland

False wetland

False dryland

FIGURE 1.12
Traditional decision tree versus spatial decision tree output for classifying data from satellite imagery. (a) Prediction of decision tree, (b) Prediction of spatial decision tree, and (c) Legend of prediction map. (From Jiang, Z et al. 2013. In *2013 IEEE 13th International Conference on Data Mining*, 320–329. doi:10.1109/ICDM.2013.96.)

(NDVI) for the years 2003, 2005, and 2008. Figure 1.12a shows the output of the traditional decision tree algorithm. The legend of the prediction maps is shown in Figure 1.12c. The green and red colors represent correctly classified wetland and correctly classified dryland. The black and blue colors represent false wetland and false dryland.

As shown in Figure 1.12a, the prediction of the traditional decision tree model has lots of salt-and-pepper noise due to high local variation of features within patches of the same class. For example, the area in the yellow circle is a dryland area consisting of trees. The black salt-and-pepper noise pixels inside the yellow circle correspond to locations without tree coverage. These pixels are misclassified as wetland here due to the i.i.d. assumption. In contrast, the spatial decision tree employs a model where the tree traversal for a location is based on not only local but also focal (i.e., neighborhood) properties of the location, thus accounting for spatial autocorrelation. Hence, as shown in Figure 1.12b, the spatial decision tree model captures the local variations results in much less salt-and-pepper noise in the same area.

The spatial data mining literature includes spatial hotspot analysis (Kulldorff 1997, 1999; E. Eftelioglu et al. 2014), discovering spatial co-location and co-occurrence patterns (Huang, Shekhar, and Xiong 2004; M. Celik et al. 2006a,b; Yoo et al. 2006; P. Mohan et al. 2010, 2011, 2012), network summarization (Oliver et al. 2010, 2014; Evans et al. 2012), GPS track mining (Fu, Hu, and Tan 2005; Sacharidis et al. 2008; Won et al. 2009; Li et al. 2010; Chen et al. 2011; W. Liu et al. 2011; Min and Wynter 2011; Yuan et al. 2011; D. Zhang et al.

2011; Y. Zheng and Zhou 2011; K. Zheng et al. 2012), spatial outlier detection (Shekhar, Lu, and Zhang 2001, 2003), spatial classification and regression (Kazar et al. 2004; Z. Jiang et al. 2012, 2015), and change footprint detection (Zhou, Shekhar, and Ali 2014).

1.5 Conclusion

The specific properties of geospatial data; its volume, variety, and velocity; and the implicit but complex nature of spatial relationships are nontrivial considerations in all geo-related research. We believe the current practice of independent research in siloed fields is counterproductive and likely untenable in the long term. We are proposing therefore that statistics, mathematics, and computer science all be considered integral to geospatial data science. This chapter explored the emerging field of geospatial data science from such a transdisciplinary perspective where these three closely related scientific disciplines are considered as integral parts of geospatial data science rather than individual siloed disciplines. Our proposed definition aims to reduce the redundant work being done across silos and to understand the limits of geospatial data science.

In the future, we envision that geospatial data science will accomplish its tasks while addressing users' privacy and confidentiality concerns. In addition, there are other issues that will need to be considered such as "the trade-offs across disciplines"; "when to use high-dimensional tools and approaches for geospatial datasets"; "how to apply spatial statistics, which assumes isotropic Euclidean space, on geospatial network datasets (e.g., road networks affect isotropy in space)"; and "how to determine the statistical distribution of geospatial datasets (e.g., GPS trajectories) in a study area." Finally, predictability and prediction error bounds should be considered since these will provide confidence limits to future approaches of geospatial data science.

References

Abrahams, P. 1987. What is computer science? *Communications of the ACM* 30 (6): 472–473. doi:10.1145/214762.315731.

Agarwal, D., A. McGregor, J. M. Phillips, S. Venkatasubramanian, and Z. Zhu. 2006. Spatial scan statistics: Approximations and performance study. In *Proceedings of the 12th ACM SIGKDD International Conference on Knowledge Discovery and Data Mining*, 24–33. ACM.

Agrawal, R., R. Srikant et al. 1994. Fast algorithms for mining association rules. In *Proceedings of the 20th International Conference on Very Large Data Bases, VLDB*, 1215:487–499. https://www.it.uu.se/edu/course/homepage/infoutv/ht08/vldb94_rj.pdf.

Aji, A., F. Wang, H. Vo, R. Lee, Q. Liu, X. Zhang, and J. Saltz. 2013. Hadoop GIS: A high performance spatial data warehousing system over mapreduce. *Proceedings of the VLDB Endowment* 6 (11): 1009–1020. doi:10.14778/2536222.2536227.

Ali, E., and M. Mokbel. 2017. SpatialHadoop. Accessed on March 12. http://spatial-hadoop.cs.umn.edu/.

Apache Spark™. 2017. Apache Spark™—Lightning-Fast Cluster Computing. Accessed on March 12. http://spark.apache.org/.

Ayres, R. U. 1997. *Information, Entropy, and Progress: A New Evolutionary Paradigm.* Springer Science & Business Media. https://books.google.com/books?hl=en&lr=&id=AgijQrntlzYC&oi=fnd&pg=PR9&dq=Information,+entropy,+and+progress:+a+new+evolutionary+paradigm&ots=KPM72Atnwu&sig=t8qiaJqJok8rw8lkP9CfusTv3EY.

Babyak, M. A. 2004. What you see may not be what you get: A brief, nontechnical introduction to overfitting in regression-type models. *Psychosomatic Medicine* 66 (3): 411–421.

Barthelemy, M. 2011. Spatial networks. *Physics Reports* 499 (1–3): 1–101. doi:10.1016/j.physrep.2010.11.002.

Borthakur, D. 2007. The Hadoop distributed file system: Architecture and design. *Hadoop Project Website* 11 (2007): 21.

Bottou, L. 2010. Large-scale machine learning with stochastic gradient descent. In *Proceedings of COMPSTAT'2010*, 177–186. Springer. http://link.springer.com/chapter/10.1007/978-3-7908-2604-3_16.

Boyd, S., N. Parikh, E. Chu, B. Peleato, and J. Eckstein. 2011. Distributed optimization and statistical learning via the alternating direction method of multipliers. *Foundations and Trends® in Machine Learning* 3 (1): 1–122.

Brewer, E. A. 2000. Towards robust distributed systems. In *PODC*. Vol. 7. http://awoc.wolski.fi/dlib/big-data/Brewer_podc_keynote_2000.pdf.

Brimberg, J., P. Hansen, K.-W. Lin, N. Mladenović, and M. Breton. 2003. An oil pipeline design problem. *Operations Research* 51 (2): 228–239.

Broder, A., R. Kumar, F. Maghoul, P. Raghavan, S. Rajagopalan, R. Stata, A. Tomkins, and J. Wiener. 2000. Graph structure in the Web. *Computer Networks* 33 (1): 309–320.

Brooks, S., A. Gelman, G. L. Jones, and X.-L. Meng. 2011. *Handbook of Markov Chain Monte Carlo*. Chapman and Hall/CRC. http://www.crcnetbase.com/doi/pdf/10.1201/b10905-1.

Butler, D. 2013. When Google got flu wrong. *Nature* 494 (7436): 155–156. doi:10.1038/494155a.

Cam, L. L., and G. L. Yang. 2000. *Independent, Identically Distributed Observations*. New York: Springer, 175–239. doi:10.1007/978-1-4612-1166-2_7.

Campbell, J. B., and R. H. Wynne. 2011. *Introduction to Remote Sensing*, Fifth Edition. New York, NY: Guilford Press.

Celik, M., S. Shekhar, J. P. Rogers, and J. A. Shine. 2006a. Sustained emerging spatio-temporal co-occurrence pattern mining: A summary of results. In *Tools with Artificial Intelligence, 2006. ICTAI'06. 18th IEEE International Conference on*, 106–115. IEEE. http://ieeexplore.ieee.org/abstract/document/4031887/.

Celik, M., S. Shekhar, J. P. Rogers, J. A. Shine, and J. S. Yoo. 2006b. Mixed-drove spatio-temporal co-occurence pattern mining: A summary of results. In *Sixth International Conference on Data Mining (ICDM'06)*, 119–128. doi:10.1109/ICDM.2006.112.

Chang, K.-T. 2006. *Introduction to Geographic Information Systems*. McGraw-Hill Higher Education Boston. http://sutlib2.sut.ac.th/sut_contents/H131376.pdf.

Chen, C., D. Zhang, P. S. Castro, N. Li, L. Sun, and S. Li. 2011. Real-time detection of anomalous taxi trajectories from GPS traces. In *Mobile and Ubiquitous Systems: Computing, Networking, and Services*, edited by A. Puiatti and T. Gu, 63–74. Lecture notes of the Institute for Computer Sciences, Social Informatics and Telecommunications Engineering 104. Springer Berlin Heidelberg. doi:10.1007/978-3-642-30973-1_6.

Chiles, J.-P., and P. Delfiner. 2012. Geostatistics: Modeling Spatial Uncertainty. Hoboken, NJ: John Wiley & Sons.

Clementini, E., and P. D. Felice. 1996. An algebraic model for spatial objects with indeterminate boundaries. *Geographic Objects with Indeterminate Boundaries* 2: 155–169.

Cliff, A. D., and J. K. Ord. 1981. *Spatial Processes: Models and Applications*. London, UK: Pion.

Codd, E. F. 1970. A relational model of data for large shared data banks. *Communications of the ACM* 13 (6): 377–387.

Cohn, A. G., and N. M. Gotts. 1996. The "egg-Yolk" representation of regions with indeterminate boundaries. *Geographic Objects with Indeterminate Boundaries* 2: 171–187.

Cohn, A. G., D. A. Randell, and Z. Cui. 1995. Taxonomies of logically defined qualitative spatial relations. *International Journal of Human-Computer Studies* 43 (5–6): 831–846.

Computational Complexity Theory. 2017. *Wikipedia*. https://en.wikipedia.org/w/index.php?title=Computational_complexity_theory&oldid=765537903.

Cormen, T. H. 2009. *Introduction to Algorithms*. MIT Press. https://books.google.com/books?hl=en&lr=&id=aefUBQAAQBAJ&oi=fnd&pg=PR5&dq=Introduction+to+algorithms&ots=dMawRuULa2&sig=HbXrvcPRnN3PObymDDm6S_ipzNc.

Curran-Everett, D. 2009. Explorations in statistics: Confidence intervals. *AJP: Advances in Physiology Education* 33 (2): 87–90. doi:10.1152/advan.00006.2009.

Daston, L., and P. Galison. 2007. *Objectivity*. Boston, MA: Zone Books.

Dean, J., and S. Ghemawat. 2008. Mapreduce: Simplified data processing on large clusters. *Communications of the ACM* 51 (1): 107–113.

Dempster, A. P., N. M. Laird, and D. B. Rubin. 1977. Maximum likelihood from incomplete data via the EM algorithm. *Journal of the Royal Statistical Society. Series B (Methodological)*, 1–38.

Dijkstra, E. W. 1959. A note on two problems in connexion with graphs. *Numerische Mathematik* 1 (1): 269–271.

Dixon, P. 2002. Ripley's K-function. In *The Encyclopedia of Environmetrics*, edited by A. H. El-Shaarawi and W. W. Piergorsch, pp. 1976–1803. New York, NY: John Wiley & Sons.

Douglas-Mankin, K. R., R. Srinivasan, and J. G. Arnold. 2010. Soil and water assessment tool (SWAT) model: Current developments and applications. *Transactions of the Asabe* 53 (5): 1423–1431.

Drineas, P., and X. Huo. 2016. *Executive Summary of Workshop on "Theoretical Foundations of Data Science (TFoDS)".*

Drummond, C. 2009. Replicability is not reproducibility: Nor is it good science. http://cogprints.org/7691/.

Dueker, K. J., and R. Vrana. 1992. Dynamic segmentation revisited: A milepoint linear data model. *Journal of the Urban and Regional Information Systems Association* 4 (2): 94–105.

Eftelioglu, E., Z. Jiang, R. Ali, and S. Shekhar. 2016a. Spatial computing perspective on food energy and water nexus. *Journal of Environmental Studies and Sciences* 6 (1): 62–76.

Eftelioglu, E., Y. Li, X. Tang, S. Shekhar, J. M. Kang, and C. Farah. 2016b. Mining network hotspots with holes: A summary of results. In *Geographic Information Science*, edited by J. A. Miller, D. O'Sullivan, and N. Wiegand, 51–67. Lecture notes in Computer Science 9927. Springer International Publishing. doi:10.1007/978-3-319-45738-3_4.

Eftelioglu, E., S. Shekhar, J. M. Kang, and C. C. Farah. 2016c. Ring-shaped hotspot detection. *IEEE Transactions on Knowledge and Data Engineering* 28 (12): 3367–3381. doi:10.1109/TKDE.2016.2607202.

Eftelioglu, E., S. Shekhar, D. Oliver, X. Zhou, M. R. Evans, Y. Xie, J. M. Kang, R. Laubscher, and C. Farah. 2014. Ring-shaped hotspot detection: A summary of results. In *2014 IEEE International Conference on Data Mining*, 815–820. doi:10.1109/ICDM.2014.13.

Eftelioglu, E., X. Tang, and S. Shekhar. 2015. Geographically robust hotspot detection: A summary of results. In *Data Mining Workshop (ICDMW), 2015 IEEE International Conference on*, 1447–1456. IEEE. http://ieeexplore.ieee.org/abstract/document/7395840/.

Ehlschlaeger, C. R., and M. F. Goodchild. 1994. Uncertainty in spatial data: Defining, visualizing, and managing data errors. In *Proceedings of GIS/LIS*, 246–253.

Elmasri, R., and S. B. Navathe. 2015. *Fundamentals of Database Systems*. 7 edition. Hoboken, NJ: Pearson.

Es, H. M. Van, C. P. Gomes, M. Sellmann, and C. L. Van Es. 2007. Spatially-balanced complete block designs for field experiments. doi:10.1016/j.geoderma.2007. 04.017.

Evans, M. R., D. Oliver, S. Shekhar, and F. Harvey. 2012. Summarizing trajectories into K-primary corridors: A summary of results. In *Proceedings of the 20th International Conference on Advances in Geographic Information Systems*, 454–457. SIGSPATIAL '12. New York: ACM. doi:10.1145/2424321.2424388.

Evans, M. R., D. Oliver, K. S. Yang, X. Zhou, and S. Shekhar. 2013. Enabling spatial big data via CyberGIS: Challenges and opportunities. *CyberGIS: Fostering a New Wave of Geospatial Innovation and Discovery. Springer Book.* https://pdfs.semanticscholar.org/a831/2a5a1ffca6a05e4788d35e81fa8d1b9ee43d.pdf.

Evans, M. R., D. Oliver, X. Zhou, and S. Shekhar. 2014. Spatial big data. *Big Data: Techniques and Technologies in Geoinformatics*, 149–176.

Freeman, L. C. 1978. Centrality in social networks conceptual clarification. *Social Networks* 1 (3): 215–239.

Fu, Z., W. Hu, and T. Tan. 2005. Similarity based vehicle trajectory clustering and anomaly detection. In *Image Processing, 2005. ICIP 2005. IEEE International Conference on*, 2:II–602. IEEE. http://ieeexplore.ieee.org/abstract/document/1530127/.

Gabriel, K. R., and J. Neumann. 1962. A Markov chain model for daily rainfall occurrence at Tel Aviv. *Quarterly Journal of the Royal Meteorological Society* 88 (375): 90–95. doi:10.1002/qj.49708837511.

Gassman, P. W., M. R. Reyes, C. H. Green, and J. G. Arnold. 2007. The soil and water assessment tool: Historical development, applications, and future research directions invited review series. *Transactions of the American Society of Agricultural and Biological Engineers* 50 (4): 1211–1250.

Gauch, H. G. 2003. *Scientific Method in Practice.* New York, NY: Cambridge University Press.

Geisberger, R., P. Sanders, D. Schultes, and D. Delling. 2008. Contraction hierarchies: Faster and simpler hierarchical routing in road networks. In *International Workshop on Experimental and Efficient Algorithms*, 319–333. Springer. http://link. springer.com/chapter/10.1007/978-3-540-68552-4_24.

Getis, A., and J. K. Ord. 2010. The analysis of spatial association by use of distance statistics. *Geographical Analysis* 24 (3). Blackwell Publishing Ltd: 189–206. doi: 10.1111/j.1538-4632.1992.tb00261.x.

Ghemawat, S., H. Gobioff, and S.-T. Leung. 2003. The google file system. In *ACM SIGOPS Operating Systems Review*, 37:29–43. ACM. http://dl.acm.org/citation. cfm?id=945450.

Giacinto, G., F. Roli, and L. Bruzzone. 2000. Combination of neural and statistical algorithms for supervised classification of remote-sensing images. *Pattern Recognition Letters* 21 (5): 385–397.

Gislason, P. O., J. A. Benediktsson, and J. R. Sveinsson. 2006. Random forests for land cover classification. *Pattern Recognition Letters* 27 (4): 294–300.

Golub, G. H., and C. Reinsch. 1970. Singular value decomposition and least squares solutions. *Numerische Mathematik* 14 (5): 403–420.

Govan, A. 2006. Introduction to optimization. In *North Carolina State University, SAMSI NDHS, Undergraduate Workshop*. https://www.ncsu.edu/crsc/events/ ugw06/presentations/aygovan/OptimizationUW06.pdf.

Graumann, A., T. Houston, J. Lawrimore, D. Levinson, N. Lott, S. McCown, S. Stephens, and D. Wuertz. 2005. Hurricane Katrina: A climatological perspective. *NOAA National Climate Data Center Technical Report 1*.

Guggenheim, E. A. 1985. Thermodynamics—An advanced treatment for chemists and physicists. *Amsterdam, North-Holland, 1985, 414 P.* http://adsabs.harvard. edu/abs/1985anh..book.....G.

Guttman, A. 1984. *R-Trees: A Dynamic Index Structure for Spatial Searching.* Vol. 14. 2. ACM. http://dl.acm.org/citation.cfm?id=602266.

Hall, M. A., and G. Holmes. 2003. Benchmarking attribute selection techniques for discrete class data mining. *IEEE Transactions on Knowledge and Data Engineering* 15 (6): 1437–1447.

Han, J, J. Pei, and Y. Yin. 2000. Mining frequent patterns without candidate generation. In *ACM Sigmod Record*, 29:1–12. ACM. http://dl.acm.org/citation.cfm?id= 335372.

Härdle, W. K., S. Klinke, and B. Röonz. 2015. Sampling theory. In *Introduction to Statistics*, Cham: Springer International Publishing. pp. 209–249. doi: 10.1007/978-3-319-17704-5_7.

Hartigan, J. A., and M. A. Wong. 1979. Algorithm AS 136: A K-means clustering algorithm. *Journal of the Royal Statistical Society. Series C (Applied Statistics)* 28 (1): 100–108.

Hong, H., B. P. Carlin, T. A. Shamliyan, J. F. Wyman, R. Ramakrishnan, F. Sainfort, and R. L. Kane. 2013. Comparing Bayesian and frequentist approaches for multiple outcome mixed treatment comparisons. *Medical Decision Making* 33 (5). SAGE PublicationsSage CA: Los Angeles, CA: 702–714. doi:10.1177/0272989X13481110.

Huang, Y., S. Shekhar, and H. Xiong. 2004. Discovering colocation patterns from spatial data sets: A general approach. *IEEE Transactions on Knowledge and Data Engineering* 16 (12): 1472–1485.

Intel. 2013. Intel distribution for Apache Hadoop software. http://www.intel.com/content/dam/www/public/us/en/documents/articles/intel-distribution-for-apache-hadoop-product-brief.pdf.

Jiang, Z., S. Shekhar, P. Mohan, J. Knight, and J. Corcoran. 2012. Learning spatial decision tree for geographical classification: A summary of results. In *Proceedings of the 20th International Conference on Advances in Geographic Information Systems*, 390–393. ACM. http://dl.acm.org/citation.cfm?id=2424372.

Jiang, Z., S. Shekhar, X. Zhou, J. Knight, and J. Corcoran. 2013. Focal-test-based spatial decision tree learning: A summary of results. In *2013 IEEE 13th International Conference on Data Mining*, 320–329. doi:10.1109/ICDM.2013.96.

Jiang, Z., S. Shekhar, X. Zhou, J. Knight, and J. Corcoran. 2015. Focal-test-based spatial decision tree learning. *IEEE Transactions on Knowledge and Data Engineering* 27 (6): 1547–1559. doi:10.1109/TKDE.2014.2373383.

Johansen, S. 1991. Estimation and hypothesis testing of cointegration vectors in Gaussian vector autoregressive models. *Econometrica: Journal of the Econometric Society* 59: 1551–1580.

Jolliffe, I. 2002. *Principal Component Analysis*. Wiley Online Library. http://onlinelibrary.wiley.com/doi/10.1002/9781118445112.stat06472/full.

Karimi, H. A. 2014. *Big Data: Techniques and Technologies in Geoinformatics*. CRC Press. http://www.crcnetbase.com/doi/pdf/10.1201/b16524-1.

Kazar, B. M., S. Shekhar, D. J. Lilja, and D. Boley. 2004. A parallel formulation of the spatial auto-regression model for mining large geo-spatial datasets. In *SIAM International Conference on Data Mining Workshop on High Performance and Distributed Mining (HPDM2004)*. Vol. 72. Citeseer. http://citeseerx.ist.psu.edu/viewdoc/download?doi=10.1.1.75.1861&rep=rep1&type=pdf.

Kleinberg, J. M., R. Kumar, P. Raghavan, S. Rajagopalan, and A. S. Tomkins. 1999. The Web as a graph: Measurements, models, and methods. In *International Computing and Combinatorics Conference*, 1–17. Springer. http://link.springer.com/chapter/10.1007/3-540-48686-0_1.

Köhler, E., K. Langkau, and M. Skutella. 2002. Time-expanded graphs for flow-dependent transit times. In *Algorithms—ESA 2002*, edited by R. Möhring and R. Raman, 599–611. Lecture notes in Computer Science 2461. Berlin Heidelberg: Springer. doi:10.1007/3-540-45749-6_53.

Kulldorff, M. 1997. A spatial scan statistic. *Communications in Statistics—Theory and Methods* 26 (6). Marcel Dekker, Inc.: 1481–1496. doi:10.1080/03610929708831995.

Kulldorff, M. 1999. Spatial scan statistics: Models, calculations, and applications. In *Scan Statistics and Applications*, 303–322. Springer. http://link.springer.com/chapter/10.1007/978-1-4612-1578-3_14.

Kwan, M.-P., I. Casas, and B. Schmitz. 2004. Protection of geoprivacy and accuracy of spatial information: How effective are geographical masks? *Cartographica: The International Journal for Geographic Information and Geovisualization* 39 (2): 15–28.

Lazer, D., R. Kennedy, G. King, and A. Vespignani. 2014. The parable of Google flu: Traps in big data analysis. *Science* 343 (6176): 1203–1205. doi:10.1126/science.1248506.

Lee, J.-H., Y.-K. Lee, K.-Y. Whang, and I.-Y. Song. 1997. A physical database design method for multidimensional file organizations. *Information Sciences* 102 (1–4): 31–65.

Legendre, P, M. R. T. Dale, M.-J. Fortin, P. Casgrain, J. Gurevitch, M.-J. Fortin, P. Ca, and J. Gurevitch4. 2004. Effects of spatial structures on the results of field experiments. *Source: Ecology Ecology* 85 (12): 3202–3214.

Li, Z., M. Ji, J.-G. Lee, L.-A. Tang, Y. Yu, J. Han, and R. Kays. 2010. MoveMine: Mining moving object databases. In *Proceedings of the 2010 ACM SIGMOD International Conference on Management of Data*, 1203–1206. ACM. http://dl.acm.org/citation.cfm?id=1807319.

Lillesand, T., R. W. Kiefer, and J. Chipman. 2014. *Remote Sensing and Image Interpretation*. Hoboken, NJ: John Wiley & Sons.

Li-Ping, C., W. Ru, S. Hang, X. Xin-Ping, Z. Jin-Song, L. Wei, and C. Xu. 2003. Structural properties of US flight network. *Chinese Physics Letters* 20 (8): 1393.

Liu, L.-K., and E. Feig. 1996. A block-based gradient descent search algorithm for block motion estimation in video coding. *IEEE Transactions on Circuits and Systems for Video Technology* 6 (4): 419–422.

Liu, W, Y. Zheng, S. Chawla, J. Yuan, and X. Xing. 2011. Discovering spatio-temporal causal interactions in traffic data streams. In *Proceedings of the 17th ACM SIGKDD International Conference on Knowledge Discovery and Data Mining*, 1010–1018. ACM. http://dl.acm.org/citation.cfm?id=2020571.

Lovell, J. 2007. Left-hand-turn elimination. *The New York Times*, December 9. http://www.nytimes.com/2007/12/09/magazine/09left-handturn.html.

Marcus, G., and E. Davis. 2014. Eight (No, Nine!) problems with big data. *The New York Times*, April 6. http://www.nytimes.com/2014/04/07/opinion/eight-no-nine-problems-with-big-data.html.

Mason, L., J. Baxter, P. L. Bartlett, and M. R. Frean. 1999. Boosting algorithms as gradient descent. In *NIPS*, 512–518. https://papers.nips.cc/paper/1766-boosting-algorithms-as-gradient-descent.pdf.

Mazzocchi, F. 2015. Could big data be the end of theory in science? *EMBO Reports*, e201541001.

McBratney, A., B. Whelan, T. Ancev, and J. Bouma. 2005. Future directions of precision agriculture. *Precision Agriculture* 6 (1): 7–23.

Melgani, F., and L. Bruzzone. 2004. Classification of hyperspectral remote sensing images with support vector machines. *IEEE Transactions on Geoscience and Remote Sensing* 42 (8): 1778–1790.

Miller, H. J., and J. Han. 2009. *Geographic Data Mining and Knowledge Discovery*. New York: CRC Press.

Miller, K. S., and B. Ross. 1993. An introduction to the fractional calculus and fractional differential equations. http://www.citeulike.org/group/14583/article/4204050.

Min, W., and L. Wynter. 2011. Real-time road traffic prediction with spatio-temporal correlations. *Transportation Research Part C: Emerging Technologies* 19 (4): 606–616.

Mislove, A., M. Marcon, K. P. Gummadi, P. Druschel, and B. Bhattacharjee. 2007. Measurement and analysis of online social networks. In *Proceedings of the 7th ACM SIGCOMM Conference on Internet Measurement*, 29–42. ACM. http://dl.acm.org/citation.cfm?id=1298311.

Mohan, P., S. Shekhar, J. A. Shine, and J. P. Rogers. 2010. Cascading spatio-tempo-ral pattern discovery: A summary of results. In *Proceedings of the 2010 SIAM International Conference on Data Mining*, 327–338. SIAM. http://epubs.siam.org/doi/abs/10.1137/1.9781611972801.29.

Mohan, P., S. Shekhar, J. A. Shine, and J. P. Rogers. 2012. Cascading spatio-temporal pattern discovery. *IEEE Transactions on Knowledge and Data Engineering* 24 (11): 1977–1992. doi:10.1109/TKDE.2011.146.

Mohan, P., S. Shekhar, J. A. Shine, J. P. Rogers, Z. Jiang, and N. Wayant. 2011. A neigh-borhood graph based approach to regional co-location pattern discovery: A summary of results. In *Proceedings of the 19th ACM SIGSPATIAL International Conference on Advances in Geographic Information Systems*, 122–132. GIS '11. New York, NY: ACM. doi:10.1145/2093973.2093991.

Møller, J., and R. P. Waagepetersen. 2007. Modern statistics for spatial point pro-cesses. *Scandinavian Journal of Statistics* 34 (4): 643–684.

National Hurricane Center. 2017. *National Hurricane Center*. Accessed March 11. http://www.nhc.noaa.gov/.

Neill, D. B., and H. J. Heinz. 2009. Expectation-based scan statistics for monitor-ing spatial time series data. *International Journal of Forecasting* 25: 498–517. doi:10.1016/j.ijforecast.2008.12.002.

Neter, J., M. H. Kutner, C. J. Nachtsheim, and W. Wasserman. 1996. *Applied Linear Statistical Models*. Vol. 4. Irwin Chicago. https://mubert.marshall.edu/bert/syl-labi/328620150114089905097122.pdf.

Newman, W. R., and L. M. Principe. 1998. Alchemy Vs. Chemistry: The etymological origins of a Historiographic mistake1. *Early Science and Medicine* 3 (1): 32–65. doi:10.1163/157338298X00022.

NSF Workshop on Geospatial Data Science. 2017. Accessed March 12. http://cybergis.illinois.edu/events/geodatascience_workshop/home.

Oliver, D., A. Bannur, J. M. Kang, S. Shekhar, and R. Bousselaire. 2010. A K-main routes approach to spatial network activity summarization: A summary of results. In *2010 IEEE International Conference on Data Mining Workshops*, 265–272. doi:10.1109/ICDMW.2010.156.

Oliver, D., S. Shekhar, J. M. Kang, R. Laubscher, V. Carlan, and A. Bannur. 2014. A K-main routes approach to spatial network activity summarization. *IEEE Transactions on Knowledge and Data Engineering* 26 (6): 1464–1478. doi:10.1109/TKDE.2013.135.

Page, L., S. Brin, R. Motwani, and T. Winograd. 1999. The pagerank citation rank-ing: Bringing order to the Web. Stanford InfoLab. http://ilpubs.stanford.edu:8090/422.

Pan, J.-X., and K.-T. Fang. 2002. *Maximum Likelihood Estimation*. New York: Springer. 77–158. doi:10.1007/978-0-387-21812-0_3.

Pang, L. X., S. Chawla, B. Scholz, and G. Wilcox. 2013. A scalable approach for LRT computation in GPGPU environments. In *Asia-Pacific Web Conference*, 595–608. Springer. http://link.springer.com/chapter/10.1007/978-3-642-37401-2_58.

Papadimitriou, C. H. 2003. *Computational Complexity*. John Wiley and Sons Ltd. http://dl.acm.org/citation.cfm?id=1074233.

Peng, R. D. 2011. Reproducible research in computational science. *Science* 334 (6060): 1226–1227. doi:10.1126/science.1213847.

Planet.gpx—OpenStreetMap Wiki. 2017. Accessed on March 12. http://wiki.open-streetmap.org/wiki/Planet.gpx.

Prasad, S. K., S. Shekhar, M. McDermott, X. Zhou, M. Evans, and S. Puri. 2013. GPGPU—Accelerated interesting interval discovery and other computations on geospatial datasets: A summary of results. In *Proceedings of the 2Nd ACM SIGSPATIAL International Workshop on Analytics for Big Geospatial Data*, 65–72. BigSpatial '13. New York, NY: ACM. doi:10.1145/2534921.2535837.

Quinlan, J. R. 1986. Induction of decision trees. *Machine Learning* 1 (1): 81–106.

Randell, D. A., and A. G. Cohn. 1992. Exploiting lattices in a theory of space and time. *Computers and Mathematics with Applications* 23 (6–9): 459–476.

Reed, D. A., and J. Dongarra. 2015. Exascale computing and big data. *Communications of the ACM* 58 (7): 56–68.

Rice, J. A. 1995. *Mathematical Statistics and Data Analysis*. Belmont, CA: Duxbury Press.

Sacharidis, D., K. Patroumpas, M. Terrovitis, V. Kantere, M. Potamias, K. Mouratidis, and T. Sellis. 2008. On-line discovery of hot motion paths. In *Proceedings of the 11th International Conference on Extending Database Technology: Advances in Database Technology*, 392–403. ACM. http://dl.acm.org/citation.cfm?id=1353392.

Safavian, S. R., and D. Landgrebe. 1991. A survey of decision tree classifier methodology. *IEEE Transactions on Systems, Man, and Cybernetics* 21 (3): 660–674.

Samet, H. 1990. *The Design and Analysis of Spatial Data Structures*. Vol. 199. Reading, MA: Addison-Wesley. http://books.google.com/books/about/The_Design_and_Analysis_of_Spatial_Data.html?id=LttQAAAAMAAJ.

Samet, H. 2015. Sorting Spatial Data. http://www.cs.umd.edu/~hjs/pubs/geoencycl.pdf.

Shekhar, S., and S. Chawla. 2003. *Spatial Databases: A Tour*. 1 edition. Upper Saddle River, NJ: Prentice Hall.

Shekhar, S., S. Chawla, S. Ravada, A. Fetterer, X. Liu, and C.-T. Lu. 1999. Spatial databases—Accomplishments and research needs. *IEEE Transactions on Knowledge and Data Engineering* 11 (1): 45–55.

Shekhar, S., M. R. Evans, J. M. Kang, and P. Mohan. 2011. Identifying patterns in spatial information: A survey of methods. *Wiley Interdisciplinary Reviews: Data Mining and Knowledge Discovery* 1 (3): 193–214. doi:10.1002/widm.25.

Shekhar, S., S. K. Feiner, and W. G. Aref. 2015. Spatial computing. *Communication of the ACM* 59 (1): 72–81. doi:10.1145/2756547.

Shekhar, S., Z. Jiang, R. Y. Ali, E. Eftelioglu, X. Tang, V. M. V. Gunturi, and X. Zhou. 2015. Spatiotemporal data mining: A computational perspective. *ISPRS International Journal of Geo-Information* 4 (4): 2306–2338. doi:10.3390/ijgi4042306.

Shekhar, S., C.-T. Lu, and P. Zhang. 2001. Detecting graph-based spatial outliers: Algorithms and applications (a summary of results). In *Proceedings of the Seventh ACM SIGKDD International Conference on Knowledge Discovery and Data Mining*, 371–376. ACM. http://dl.acm.org/citation.cfm?id=502567.

Shekhar, S., C.-T. Lu, and P. Zhang. 2003. A unified approach to detecting spatial outliers. *GeoInformatica* 7 (2): 139–166.

Silver, M. R., and O. L. De Weck. 2007. Time-expanded decision networks: A framework for designing evolvable complex systems. *Systems Engineering* 10 (2): 167–188.

Stein, A., and L. C. A. Corsten. 1991. Universal kriging and cokriging as a regression procedure. *Biometrics* 47 (2): 575. doi:10.2307/2532147.

Stolorz, P. E., H. Nakamura et al. 1995. Fast spatio-temporal data mining of large geophysical Datasets. In *Proceedings of the First International Conference on Knowledge Discovery and Data Mining (KDD-95)*, Montreal, Canada, August 20–21, 1995, edited by U. M. Fayyad and R. Uthurusamy, pp. 300–305. AAAI Press. http://www.aaai.org/Library/KDD/1995/kdd95-035.php.

Taleb, N. N. 2007. *The Black Swan: The Impact of the Highly Improbable*. Vol. 2. Random House. https://books.google.com/books?hl=en&lr=&id=gWW4SkJjM08C&oi=fnd&pg=PR33&dq=The+black+swan:+The+impact+of+the+highly+improbable&ots=v-zQIWTLgw&sig=LlVA4ILx8dlKuH6V48qk_vF7rDY.

Tang, X., E. Eftelioglu, D. Oliver, and S. Shekhar. 2017. Significant linear hotspot discovery. *IEEE Transactions on Big Data*, 1–1. doi:10.1109/TBDATA.2016.2631518.

Tang, X., E. Eftelioglu, and S. Shekhar. 2015. Elliptical hotspot detection: A summary of results. In *Proceedings of the 4th International ACM SIGSPATIAL Workshop on Analytics for Big Geospatial Data*, 15–24. BigSpatial'15. New York, NY: ACM. doi:10.1145/2835185.2835192.

Tobler, W. R. 1970. A computer movie simulating urban growth in the Detroit region. *Economic Geography* 46 (6): 234. doi:10.2307/143141.

Wall, M. M. 2004. A close look at the spatial structure implied by the CAR and SAR models. *Journal of Statistical Planning and Inference* 121: 311–324. doi:10.1016/S0378-3758(03)00111-3.

Wall, M. E., A. Rechtsteiner, and L. M. Rocha. 2003. Singular value decomposition and principal component analysis. In *A Practical Approach to Microarray Data Analysis*, 91–109. Springer. http://link.springer.com/content/pdf/10.1007/0-306-47815-3_5.pdf.

Welcome to the OGC|OGC. 2017. Accessed on March 12. http://www.opengeospatial.org/.

Wilkinson, J. H., and J. H. Wilkinson. 1965. *The Algebraic Eigenvalue Problem*. Vol. 87. Clarendon Press Oxford. http://tocs.ulb.tu-darmstadt.de/35148594.pdf.

Williams, C. K. I. 1998. Prediction with Gaussian processes: From linear regression to linear prediction and beyond. In *Learning in Graphical Models*, 599–621. Dordrecht: Springer Netherlands. doi:10.1007/978-94-011-5014-9_23.

Won, J.-I., S.-W. Kim, J.-H. Baek, and J. Lee. 2009. Trajectory clustering in road network environment. In *Computational Intelligence and Data Mining, 2009. CIDM'09. IEEE Symposium on*, 299–305. IEEE. http://ieeexplore.ieee.org/abstract/document/4938663/.

Yoo, J. S., S. Shekhar, S. Kim, and M. Celik. 2006. Discovery of co-evolving spatial event sets. In *Proceedings of the 2006 SIAM International Conference on Data Mining*, 306–315. SIAM. http://epubs.siam.org/doi/abs/10.1137/1.9781611972764.27.

Yu, J., J. Wu, and M. Sarwat. 2015. Geospark: A cluster computing framework for processing large-scale spatial data. In *Proceedings of the 23rd SIGSPATIAL International Conference on Advances in Geographic Information Systems*, 70. ACM. http://dl.acm.org/citation.cfm?id=2820860.

Yuan, J., Y. Zheng, X. Xie, and G. Sun. 2011. Driving with knowledge from the physical world. In *Proceedings of the 17th ACM SIGKDD International Conference on Knowledge Discovery and Data Mining*, 316–324. ACM. http://dl.acm.org/citation.cfm?id=2020462.

Zhang, D., N. Li, Z.-H. Zhou, C. Chen, L. Sun, and S. Li. 2011. iBAT: Detecting anomalous taxi trajectories from GPS traces. In *Proceedings of the 13th International Conference on Ubiquitous Computing*, 99–108. UbiComp '11. New York, NY: ACM. doi:10.1145/2030112.2030127.

Zhang, N., M. Wang, and N. Wang. 2002. Precision agriculture—A worldwide overview. *Computers and Electronics in Agriculture* 36 (2): 113–132.

Zheng, K., Y. Zheng, X. Xie, and X. Zhou. 2012. Reducing uncertainty of low-sampling-rate trajectories. In *Data Engineering (ICDE), 2012 IEEE 28th International Conference on*, 1144–1155. IEEE. http://ieeexplore.ieee.org/abstract/document/6228163/.

Zheng, Y., and X. Zhou. 2011. *Computing with Spatial Trajectories*. Springer Science & Business Media. https://books.google.com/books?hl=en&lr=&id=JShQJF23x BgC&oi=fnd&pg=PR3&dq=Computing+with+spatial+trajectories&ots=6M Xjft4n7_&sig=wEyPQfzxE52WQYU3WM0lqhuAU1I.

Zhou, X., S. Shekhar, and R. Y. Ali. 2014. Spatiotemporal change footprint pattern discovery: An inter-disciplinary survey. *Wiley Interdisciplinary Reviews: Data Mining and Knowledge Discovery* 4 (1): 1–23.

Zomorodian, A. 2012. Topological data analysis. *Advances in Applied and Computational Topology* 70: 1–39.

2

Geocoding Fundamentals and Associated Challenges

Claudio Owusu, Yu Lan, Minrui Zheng, Wenwu Tang, and Eric Delmelle

CONTENTS

2.1 Introduction: Geocoding and Geocoding Systems

In the twenty-first century, the ubiquitous usage of smartphones equipped with location-based services has helped millions of individuals in navigating busy traffic or finding available amenities around a particular location. Central to this technological revolution is the process of geocoding, which essentially translates text-based information about locations (address, zip code, names of localities, or even countries) into numerical geographic coordinates (e.g., longitude and latitude). Geocoding uses a spatially explicit reference dataset (e.g., digital road network) to identify the location that best matches the input address, essentially by comparing and interpolating the address to the range of addresses for each segment of the reference dataset. Each segment contains the locations of the street center and the range of addresses between the street intersections.

Geocoding is generally incorporated in commercial geographic information systems (Bichler and Balchak 2007), where geocoded data can collectively be used for mapping, visualization, and spatial analysis of events. In the past few years, however, the democratization of internet-based mapping services such as Google Maps or MapQuest has facilitated the use of online geocoding services for non-GIS users (Wu et al. 2005; Roongpiboonsopit and Karimi 2010a).

2.1.1 Applications of Geocoding

There is a myriad of domains that have benefitted from geocoding. Geocoding has been a critical element for the delivery of parcels (Jung, Lee, and Chun 2006) and for emergency dispatching management (Derekenaris et al. 2001) where locating the destination in a timely manner is critical. In health studies, geocoding has been used extensively in research with geographic themes such as health disparities (Krieger, Chen et al. 2002; Rehkopf et al. 2006), accessibility to health care (Luo and Qi 2009; Delmelle et al. 2013), disease mapping (Law et al. 2004; Delmelle et al. 2013; Delmelle, Dony et al. 2014), and environmental exposure assessment (Chakraborty and Zandbergen 2007; Zandbergen 2007). In crime analysis, geocoding technology serves as one of the important procedures to obtain data for planning, monitoring, and evaluation of targeted responses to reduce crime in communities (Chainey and Ratcliffe 2013). The process is therefore seen as a means of achieving intelligence-led policing (Ratcliffe 2002; Chainey and Ratcliffe 2013). In addition, geocoding has been used in transportation studies (Park et al. 2011; Qin et al. 2013) for the purpose of planning efficient transportation systems and preventing traffic crashes.

2.1.2 Motivation

In this chapter, we explore geocoding fundamentals, and a myriad of challenging issues that are intimately associated with the procedure, such as spelling sensitivity, accuracy, efficiency, and automation. We also focus on the assessment of the impact of uncertainties related to these geocoding issues on the discovery of spatially explicit patterns. Further, we highlight the significance of geomasking, which is particularly important to preserve confidentiality and minimize the risk of success in reverse geocoding. We then conduct a discussion on web-based geocoding and its benefits, limits, and computational hurdles. We integrate alternative web-based geocoding services together with a cross-validation approach to facilitate the impact assessment of uncertainties associated with geocoding.

In the next section, we briefly describe geocoding fundamentals and illustrate the challenges experienced when attempting to geocode our sample data (see, illustrative dataset in Section 2.1.4). In Section 2.3, we discuss geocoding quality, including sources of errors and the impact of low geocoding quality on spatial analysis. Section 2.4 is devoted to the topic of web-based geocoding, which has recently received a lot of attention. In Section 2.5, we evaluate the merits of two web-based geocoding services as an alternative to commercial geocoding software. Efforts to model and visualize the errors are also presented. In Section 2.6, we address the issue of reverse geocoding, and discuss geomasking and aggregation, two techniques particularly useful to address privacy concerns. We conclude our chapter in Section 2.7 and present avenues for future research.

2.1.3 Contributions

Besides describing and illustrating the process of geocoding, this chapter makes a series of important contributions: (1) strategies to increase the match rate for datasets that include incomplete input addresses (reengineering incomplete addresses in an effort to increase the match rate), (2) use of online geocoding services to cross-validate geocoding results obtained from commercial GIS (and estimating uncertainties in geocoding results), and (3) modeling geocoding errors.

2.1.4 Illustrative Dataset

We use a subset of historical paper records of private water well permits from Gaston County, North Carolina (from 1989 to the present, n = 7920) to illustrate the geocoding concepts (subset n = 285). Historical records were made available as part of an effort funded by the Centers for Disease Control and Prevention, aiming to establish a public digital database of the county's wells and promote the protection of private well water supplies and quality, ultimately protecting and monitoring a key portion of the county's water supply.

The dataset is particularly salient since historical records pose serious challenges such as (1) incomplete addresses or (2) paper damage. First, a complete address should have all the key components such as house number, street name, street type as well as other directional attributes when possible (e.g., 826 Union Rd, Gastonia, NC 28054). We define an address to be incomplete when any of the key components is not available in the dataset. Second, some permits have faded, making it difficult to transcribe all the address information needed for geocoding. These two problems introduce uncertainties in the datasets.

Private well permit records were scanned and information encoded in a database; each record contains information about the owner of the well, residential location, details of the parcel, ground sketch of the water well

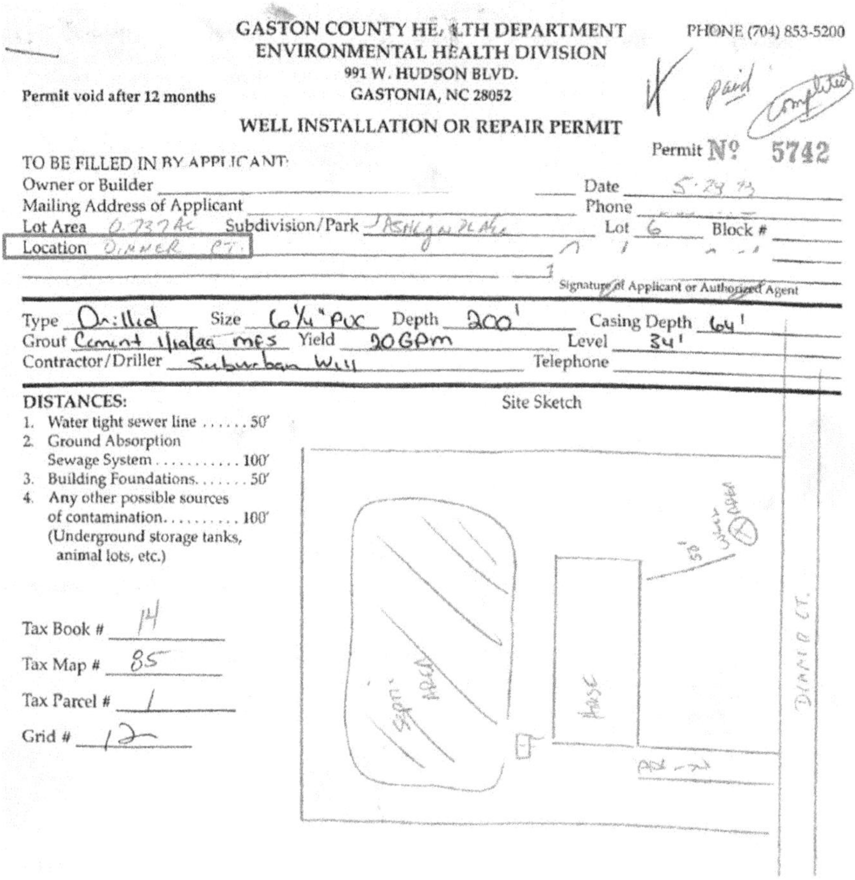
GASTON COUNTY HE, LTH DEPARTMENT
ENVIRONMENTAL HEALTH DIVISION
991 W. HUDSON BLVD.
GASTONIA, NC 28052

PHONE (704) 853-5200

Permit void after 12 months

paid Completed

WELL INSTALLATION OR REPAIR PERMIT

Permit № 5742

TO BE FILLED IN BY APPLICANT:
Owner or Builder ____ Date 5·29·93
Mailing Address of Applicant ____ Phone ____
Lot Area 0.737 Ac Subdivision/Park ____ Lot 6 Block # ____
Location Dinner Ct.

Signature of Applicant or Authorized Agent

Type Drilled Size 6¼" PVC Depth 200' Casing Depth 64'
Grout Cement ____ Yield 20 GPm Level 34'
Contractor/Driller Suburban Well Telephone ____

DISTANCES:
1. Water tight sewer line . . . . . . 50′
2. Ground Absorption Sewage System . . . . 100′
3. Building Foundations. . . . . . . 50′
4. Any other possible sources of contamination. . . . . 100′ (Underground storage tanks, animal lots, etc.)

Site Sketch

Tax Book # 14
Tax Map # 85
Tax Parcel # 1
Grid # 12

FIGURE 2.1
A typical private well permit with information of the owner (masked), location, and a sketch of where the well is built.

position, well specification, and the tax location code of the parcel. Figure 2.1 shows an example of a scanned permit. For illustration purposes, we selected a random sample of n = 285 (3%) well samples.

2.2 Geocoding Fundamentals: Input and Reference Data

Accurate reference datasets and valid addresses are the two required inputs for geocoding. *Reference datasets* typically include street network, parcel, and address points data (Zandbergen 2008). In this chapter, we use all three reference datasets and set up hierarchic rules to geocode the illustrative dataset.

Figure 2.2 shows an instance of two different reference datasets (address points and parcel centroid). It can be seen that address points reference data depicts the centroid of the buildings, making it more accurate than the other reference datasets.

For a myriad of reasons such as protecting confidentiality, *addresses* are sometimes made available at different scales, including the street level (Rushton et al. 2006; Goldberg, Wilson, and Knoblock 2007), names of buildings (Davis and Fonseca 2007), closest intersection (Levine and Kim 1998; Park et al. 2011; Delmelle, Zhu et al. 2014), neighborhood level (Casas, Delmelle, and Varela 2010), ZIP code (Krieger, Chen et al. 2002; Krieger, Waterman et al. 2002), textual descriptions of localities (Goldberg and Cockburn 2010), and cities or counties. The scale at which addresses are made available will affect the location of the output feature. For example, addresses at the ZIP code level will be geocoded at the centroid of a postal zip code instead of the residential location.

2.2.1 Geocoding Process

The geocoding process relies on a matching algorithm, which essentially attempts to determine the location of the input address over the range of addresses in the reference dataset. The reference dataset used for the

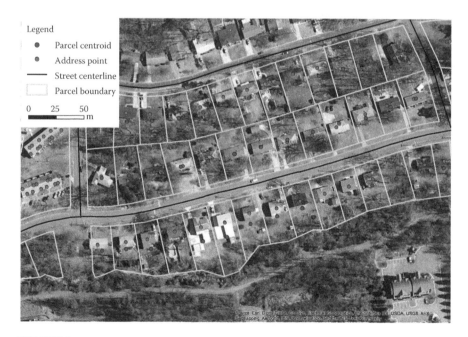

FIGURE 2.2
Example of two reference datasets: address point in red (most accurate) and parcel centroid (less accurate).

geocoding process determines the technique used in matching the spatial information to geographic coordinates. In most commercial GIS software packages, the matching algorithm is embedded in an address locator. An address locator is a model used to create geometry for textual descriptions representing addresses in the reference data (ESRI Redlands CA, USA). In the United States, a dual range address locator is used when street network is chosen as reference data.

Street geocoding is the most widely used technique due to the readily available TIGER files from the U.S. Census Bureau; here, the algorithm performs a linear interpolation of the input address within the range of address numbers and polarity of the street segment. The process can be decomposed in multiple stages. First, the algorithm attempts to match the street name of the input address with street names from the reference dataset. Next, it will determine the side of the street the address is at, based on whether the address number is even or odd. Third, the correct position of the address is determined after computation of the proportion of the address range associated with the correct side of the street segment. This proportion is then added to the start of the segment to obtain the correct coordinate. Finally, for most commercial GIS software, an optional offset from the street centerline is added. Figure 2.3 shows the interpolated distance (v) and the offset distance (d) used to determine an address along Union Road. The address range along Union Road starts from 101 to 199 on the odd parity side, and from 102 to 200 on the even parity side.

In *parcel geocoding*, the input address is matched to the centroid of the parcel. The returned geographic feature is therefore a point feature with a geographic coordinate (Zandbergen 2008). Although the technique is generally assumed to return more accurate results, it also has been found to introduce positional errors, particularly for a large parcel, since the true address location may not necessarily be at the center of that parcel.

Address point geocoding has been introduced to alleviate this problem. The input address is matched directly to a point feature, which represents the

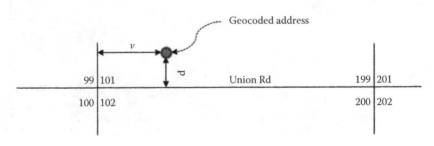

FIGURE 2.3
Interpolation algorithm using address range between the start and end of the street centerline segment for an input address as 117, Union Road.

center of the rooftop of buildings making it more accurate. Emergency calls (e.g., 911 in the United States) use such a geocoding approach.

2.2.2 Match Rate

The success of the geocoding procedure can be determined by its match rate, which is the percentage of records in the input dataset that was correctly geocoded (Zandbergen 2008). A high match rate is often desirable because geocoded results are further used as the sample during spatial investigations (Goldberg, Wilson, and Knoblock 2007; Zimmerman 2008; Ha et al. 2016). Zimmerman (2008) showed that in some instances up to 30% of addresses may need to be excluded if only geocoded records were considered during the analysis. This exclusion of unmatched records reduces the sample size, thereby weakening the generalization of the analytical results due to selection bias and reducing statistical confidence (Zimmerman 2008; Ha et al. 2016).

Geocoding is now a key research methodology and efforts to increase the match rate will help to reduce unmatched addresses that are excluded from the spatial analysis. It is important to note that an increased match rate does not automatically translate into improved geocoding quality. Different strategies exist to increase the geocoding match rate. First, *varying the spelling sensitivity* essentially increases the degree to which a street name is allowed to change. One drawback of this approach is that it will augment the set of potential matches at the cost of potentially selecting a wrong match. The second strategy consists of using *different reference datasets* (McElroy et al. 2003; Yang et al. 2004). A couple of recent studies combined parcel and street network geocoding techniques as a strategy to increase the match rate of the output geographic features (Roongpiboonsopit and Karimi 2010b; Murray et al. 2011; Delmelle et al. 2013). For instance, Delmelle et al. (2013) used different U.S. Census reference datasets to increase the number of geocoded children with birth defects in a study estimating travel impedance to health care centers.

2.2.3 Illustration

In the context of our illustrative dataset, we used multiple datasets from the Gaston County Planning & GIS Department and developed a two-phase geocoding approach as shown below in Figure 2.4. First, during an *automated phase*, different reference datasets (address point, parcel, and street network datasets) are combined in a hierarchical manner into a single composite locator in ArcGIS, a commercial GIS. The rationale to impose a hierarchy among different datasets is to increase the match rate while reducing the odds of positional error. Second, the *improvement phase* consists of using additional datasets such as bacteria test results of the wells and deed records to reengineer the unmatched addresses. Three main strategies are adopted in this

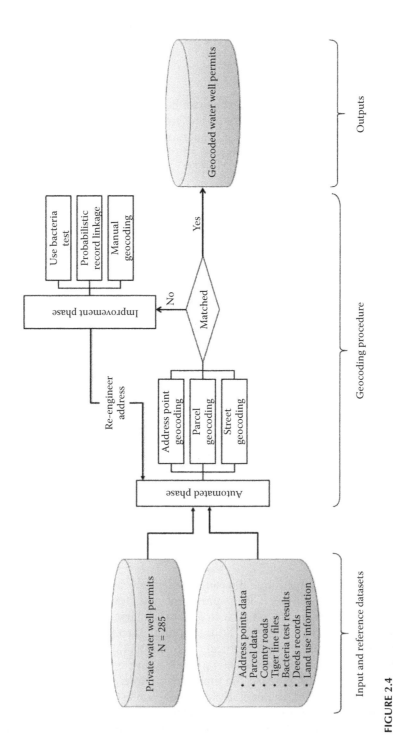

FIGURE 2.4
Flowchart: input and references datasets, geocoding procedure, and outputs.

FIGURE 2.5
N = 285 geocoded private well permits in Gaston County, North Carolina.

phase. First, the unique permit number is linked to and cross-checked with the bacteria test results. Second, non-successful records are then subject to a probabilistic record linkage, using information such as tax location codes, name of the well owner, subdivision name, lot size, and block number information. Third, manual geocoding is implemented as the final step, which involves manually interpreting the descriptive address, using additional information such as lot area, lot number, and block number. Once an address has been determined, the commercial GIS attempts to re-geocode using the composite address locator. Figure 2.5 shows the locations of the n = 285 wells that were geocoded with address points reference data.

2.3 Geocoding Quality: Sources of Errors

The success of the geocoding procedure is merely a function of the completeness of the addresses and the quality (i.e., spatial and temporal accuracy, completeness) of the local and regional road network that is used as the reference dataset (O'Reagan 1987; Krieger, Waterman et al. 2002; Zandbergen 2008; Goldberg 2011), and uncertainty with the matching algorithms (Rushton et al. 2006; Goldberg, Wilson, and Knoblock 2007; Zandbergen 2008, 2011). Over the past decades, however, the accuracy and availability of reference datasets have been improved (Dueker 1974; Werner 1974; Griffin et al. 1990; Boscoe, Ward, and Reynolds 2004).

Although street networks continue to be the most widely used referenced data, the availability of parcel datasets and the introduction of address points from emergency 911 calls in the United States have increased the accuracy and match rate (Zandbergen 2008). The input datasets have expanded from postal addresses (O'Reagan 1987) to include descriptive addresses of locations (Levine and Kim 1998; Davis and Fonseca 2007).

2.3.1 Positional Accuracy

Although the match rate indicates the percentage of addresses that are successfully geocoded, it does not inform us whether the coordinates obtained from the geocoding procedure are the true coordinates. Positional accuracy is a measure of the nearness of the geocoded output from the true location on the ground. Delmelle, Dony et al. (2014) compared geocoded cases of dengue fever in an urban environment of Colombia to locations measured from GPS devices (ground truth). In the context of our illustrative dataset, positional accuracy is estimated by comparing address points that represent the center of the rooftop of buildings with water wells obtained by geocoding from a commercial GIS.

Positional accuracy can be improved by more accurate addresses and reference datasets that are spatially and temporally accurate. Practically taking measurements with GPS devices for the events being investigated can also improve the positional accuracy, but this may be costly and timely ineffective, especially when gathering large datasets. Lastly, using alternative reference datasets for geocoding different environments may minimize the errors. For example, in rural areas where large parcels is the norm, it may be helpful to use aerial photos to generate an address point that better represents the center of the rooftop of the buildings *(if an address point dataset is not already available)* than using parcel or street network datasets.

2.3.2 Impact of Geocoding Quality

Geocoding challenges mentioned in the previous section affect the geocoding quality in terms of match rate and the positional accuracy (O'Reagan 1987; Boscoe, Ward, and Reynolds 2004). Such issues are particularly important in health studies (Bonner et al. 2003; Whitsel et al. 2004; Rushton et al. 2006; Zandbergen 2007; Mazumdar et al. 2008; Chainey and Ratcliffe 2013). Positional accuracy has been found to be critical in studies of environmental exposure as errors can lead to mischaracterization in the risk analysis (Bonner et al. 2003). Positional errors in residential addresses pose a serious challenge for spatial analysis (O'Reagan 1987; Jacquez and

Jacquez 1999; Bonner et al. 2003; Harada and Shimada 2006; Goldberg, Wilson, and Knoblock 2007; Bichler and Balchak 2007; Mazumdar et al. 2008; Zandbergen 2008; Goldberg and Cockburn 2010; Zimmerman and Li 2010; Zimmerman, Li, and Fang 2010), since it may result in (1) underestimation of local risk, (2) misplacement of high-risk areas of a disease, (3) mischaracterization in the analysis of exposure risk, (4) misevaluation of spatial association, and (5) biased evidence for decision makers. When estimating access to health care, positional errors may introduce bias in the estimation of travel impedance, especially for individuals geocoded at the ZIP code for instance.

2.4 Web-Based Geocoding

The costs to prepare reference data and standardize addresses can be prohibitive when using commercial GIS software. With the rapid development of cyber-enabled technology, a myriad of web-based providers (such as Google Maps, Bing Maps, and MapQuest, to name a few) have made the process of geocoding more accessible and faster through their online geocoding services (Roongpiboonsopit and Karimi 2010a). The preparation and maintenance of reference data, address standardization, and algorithm implementation and update for geocoding are hidden in these online services (accessible as APIs). Online geocoders typically use street network data that are more up to date, which is likely to result in lower positional errors. Online geocoders, however, have limits on the number of records that can be processed (e.g., 2500 for Google Maps and Bing Maps on a daily basis, 15,000 per month for MapQuest), suffer from a lack of transparency about the geocoding algorithm (including address interpretation) and lack of metadata on the update of reference data (an issue that may vary spatially). Another important issue is that the use of online geocoders may raise important ethical issues such as confidentiality since addresses are uploaded to remote servers. In the United States, this may violate the Health Insurance Portability and Accountability Act, which protects individuals' medical records and other personal health information (DeLuca and Kanaroglou 2015; Kirby, Delmelle, and Eberth 2017; Mak et al. 2012). Different strategies exist to circumvent this issue, such as geocoding at a coarser scale, or bundle the batch of addresses to be geocoded with random addresses (Gittler 2007; Goldberg 2008).

When using geocoding APIs, users or developers need to call functions and obtain authentication from corresponding online geocoding providers. Then these online geocoding services will use their own algorithms to calculate the coordinates that will be returned to the user (e.g., in pure text or XML-based formats). In most occasion, users can type the address that they want to geocode and click a button, to display the results on the map (i.e., in an interactive manner). Besides being available to non-GIS users, web-based geocoding systems are particularly helpful to evaluate the accuracy of the geocoding results obtained from commercial GIS software, such as ArcGIS. The accuracy evaluation is typically conducted by comparing the geocoded coordinates (Duncan et al. 2011).

2.5 Using Web-Based Geocoding Services for Cross Validation

In this study, we follow an approach similar to Duncan et al. (2011) that is based on online geocoding services (Google and MapQuest here) to validate the geocoding results from those obtained by a commercial GIS (ArcGIS). Each address record may exhibit differences in the coordinates among these geocoding options; the distance between coordinates from online geocoding services and ArcGIS-based results (referred to as error distance) is calculated. We estimate the error for the $n = 285$ geocoded samples. The distances are grouped into different "deviation categories" (<50, <100, <150, <200, <250, <300, and >300 m). For each category, we report the match rate, defined as "the percentage of the successfully geocoded records in relation to the total number of records originally subjected to the geocoding process, regardless of the positional accuracy" (Kounadi et al. 2013). Table 2.1 shows the percentage of geocoding results located in certain deviation categories according to different web-based geocoding services.

Generally, Google has a higher match rate and its geocoding results are likely to be closer to the ones obtained from ArcGIS. Depending on the purpose of the study, strict error thresholds may be necessary. In the case of studying exposure to highway pollution, a difference of 300 m may be very significant and bias the analysis (Zandbergen 2007). Further, greater distance errors are not uncommon in rural areas (Zimmerman and Li 2010). In the following section, we will analyze and model our web-based geocoding results comparing with true coordinates (in this case, we consider results of ArcGIS obtained using address point geocoding as the true coordinates).

TABLE 2.1

Variation in Match Rate for Two Online Geocoding Systems with Deviation Categories

Buffer (m)	50	100	150	200	250	300	>300
Google (%)	70.18	85.96	89.12	90.88	92.28	93.33	6.67
MapQuest (%)	62.46	82.11	89.47	90.88	92.63	92.98	7.02

2.5.1 Modeling Geocoding Error

In this study, we compare results of online geocoding services from Google and MapQuest to the ones obtained using ArcGIS. For this comparison purpose, we constructed error modeling, which consists of the following steps: (1) acquiring results from web-based geocoding services, (2) convert latitude and longitude (WGS84) into XY coordinates, (3) calculate the Euclidean error distance (in meter) between results of ArcGIS and web-based geocoding results, and (4) compare geocoding results in terms of the empirical distribution of error distance and fitted error model based on, for example, distance-decayed functions.

The error distance can be visualized in different ways. The error is represented in its simplest form as a line connecting the spatial locations of the geocoded well with the commercial solver and the online geocoders (yellow for MapQuest, red for Google) as shown in Figure 2.6a–d.

Figure 2.6e illustrates the error distance between the commercial geocoder and the Google geocoder, where a larger symbol denotes a greater error distance. Figure 2.6f compares the error distance among online providers. In pink and purple colored regions, the error distance is much lower when using Google than MapQuest, while the reverse is true for green colored regions. Figure 2.6e–f clearly suggests the presence of a spatial pattern in terms of error distance.

Table 2.2 and Figures 2.7 and 2.8 illustrate the empirical histogram and probabilistic distributions of error distance for the two web-based geocoding services (bin size: 10 m). About 95% of the Google-based results (with a median of 26.59 m) fall within a distance that is less than 250 m. MapQuest-based geocoding results (median: 39.28 m) have a longer error distance (about 360 m) than those of Google (250 m) with respect to a 95% threshold. In addition, the mode of Google-based error distance is within 10 m (covering 23.83% of the data), compared to MapQuest-based results with a mode around 30–40 m (25.62%). For the error modeling, we fitted the histograms of error distance using Pareto functions (see Morrill and Pitts 1967). Table 2.3 summarizes model fitting results. The goodness-of-fit of the error model for Google-based geocoding results (up to 88.97% of the variance explained) is much higher than that for MapQuest-based results (only 74.01% of the variance explained).

Results from both empirical distribution and the fitted error models suggest that Google's online geocoding service generally outperforms MapQuest for the geocoding task in our study area. This finding is consistent with what has been reported in the literature. For example, Roongpiboonsopit and Karimi (2010a) compared the quality of five online geocoding services (including Google and MapQuest), and found that Google provided a shorter error distance than MapQuest. Results from other relevant studies by Cui (2013), Chow et al. (2016), and Karimi et al. (2011) also indicate

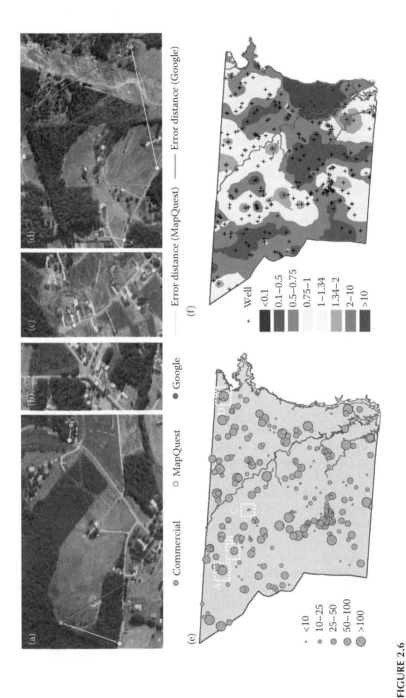

FIGURE 2.6
Visualization of a geocoding error. In a–d, geocoded addresses from the ArcGIS commercial geocoder are depicted in green, while the results from Google and MapQuest are colored in red and yellow, respectively. In e, circles of increasing sizes represent a higher error distance (units: meters). In f, is an interpolated map showing the ratio of distance errors between Google and MapQuest (in green areas, MapQuest has a lower error distance).

TABLE 2.2

Frequency and Probability of the Error Distance of Online Geocoding Services

	Google			MapQuest		
Bin	Frequency	Percent (%)	Cumulative (%)	Frequency	Percent (%)	Cumulative (%)
10	66	23.83	23.83	0	0.00	0.00
20	52	18.77	42.60	9	3.20	3.20
30	36	13.00	55.60	64	22.78	25.98
40	26	9.39	64.98	72	25.62	51.60
50	21	7.58	72.56	33	11.74	63.35
60	16	5.78	78.34	14	4.98	68.33
70	13	4.69	83.03	12	4.27	72.60
80	5	1.81	84.84	16	5.69	78.29
90	4	1.44	86.28	11	3.91	82.21
100	6	2.17	88.45	2	0.71	82.92
110	1	0.36	88.81	7	2.49	85.41
120	3	1.08	89.89	9	3.20	88.61
130	3	1.08	90.97	3	1.07	89.68
140	1	0.36	91.34	2	0.71	90.39
150	1	0.36	91.70	0	0.00	90.39
160	0	0.00	91.70	0	0.00	90.39
170	0	0.00	91.70	2	0.71	91.10
180	2	0.72	92.42	2	0.71	91.81
190	2	0.72	93.14	0	0.00	91.81
200	1	0.36	93.50	0	0.00	91.81
210	0	0.00	93.50	2	0.71	92.53
220	1	0.36	93.86	1	0.36	92.88
230	1	0.36	94.22	1	0.36	93.24
240	1	0.36	94.58	0	0.00	93.24
250	1	0.36	94.95	2	0.71	93.95
260	1	0.36	95.31	0	0.00	93.95
270	1	0.36	95.67	0	0.00	93.95
280	0	0.00	95.67	0	0.00	93.95
290	1	0.36	96.03	0	0.00	93.95
300	0	0.00	96.03	0	0.00	93.95
310	0	0.00	96.03	1	0.36	94.31
320	0	0.00	96.03	0	0.00	94.31
330	0	0.00	96.03	0	0.00	94.31
340	1	0.36	96.39	1	0.36	94.66
350	1	0.36	96.75	0	0.00	94.66
360	0	0.00	96.75	0	0.00	94.66
More	9	3.25	100.00	15	5.34	100.00

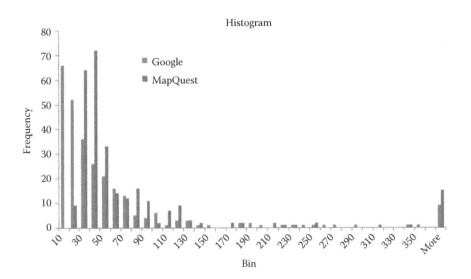

FIGURE 2.7
Histogram of error distance of online geocoding services (bin size: 10 m).

that Google's geocoding service can achieve rates that are 91.5%, 100%, and 93.64%, respectively, which are higher than other online geocoding services (e.g., MapQuest, Bing, and Geocoder.us). While multiple factors may contribute to geocoding errors, frequent update of reference data by Google may explain its high geocoding accuracy.

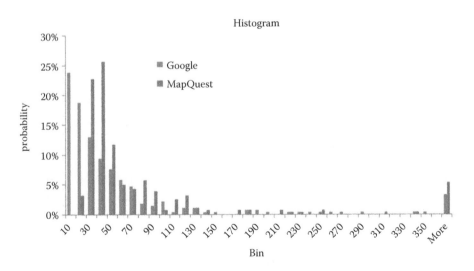

FIGURE 2.8
Empirical probabilistic distribution of error distance of online geocoding services.

TABLE 2.3

Fitted Modeling Results Based on Error Distance

Geocoding Services	Fitted Models	R²
Google	$Y = 4245\,D^{-1.508}$	0.8897
MapQuest	$Y = 6431.1\,D^{-1.523}$	0.7401

Note: Y: frequency; D: distance

2.6 Reverse Geocoding, Geomasking, and Aggregation

Although geocoded data result in a great opportunity to develop better analytical solutions, there exist some important concerns, especially in the context of epidemiology to protect privacy needs. At the core of the issue is the thread of *reverse geocoding*, which essentially determines the address based on geographic coordinates. Using a published map of geocoded records and overlaying with other layers of spatial information (such as parcel and street layers), the approximate address of the geocoded record can be traced back (Curtis, Mills, and Leitner 2006).

Several *geomasking* techniques and aggregation strategies have been developed to conceal the true identity of geocoded records and minimize the risk of success in reverse geocoding. Geomasking (Armstrong, Rushton, and Zimmerman 1999) is a spatial statistical technique used to introduce uncertainty (i.e., noise) into the spatial locations of geocoded records, which has implications for the quality of further spatial analysis (e.g., cluster detection). The main mechanism behind geomasking consists of perturbing the spatial location of a geocoded record, typically in a random distance and along a random direction. Other strategies have been developed, such as the donut geomasking method (Hampton et al. 2010) where geocoded records are moved within a random direction and within certain distance bounds. These distance bounds can be tighter in urban areas and looser in rural regions where the spacing between residences is much greater.

Finally, geocoded records can be *spatially aggregated* into census units, where all the data are moved to the geographic centroid of the unit (Tellman et al. 2010). The choice of the unit is a function of the number of cases and the population within that unit.

Despite their ability to preserve some confidentiality, *geomasking* and *aggregation* methods have some substantial limitations, such as (1) reducing the level of precision, (2) introducing statistical bias into the results, (3) blurring meaningful variations in data, and (4) weakening clustering detection. Current research attempts to find optimal geomasking strategies to preserve the spatial pattern of geocoded records while maintaining privacy.

2.7 Conclusions

In this chapter, we have discussed fundamentals of geocoding, which we illustrated on a dataset of private well addresses in Gaston County, North Carolina. We compared spatial locations estimated by a commercial geocoder to the ones obtained by two popular online providers, Google and MapQuest. We found that in most cases, coordinates from online geocoders were relatively close (26.59 m for Google and 39.28 m for MapQuest) to the ones obtained by the commercial geocoder. Generally, MapQuest geocoder yielded greater error than Google geocoder.

There remains a suite of challenges in geocoding. First, online web services provide an alternative for geocoding, but further work is needed to tackle the issue of transparency on reference datasets and geocoding algorithms. An open geocoding standard and platform may be of help. Second, massive data are increasingly available, and how to efficiently and effectively geocode these datasets (say, millions or billions of addresses) poses a big data challenge. Cyberinfrastructure-enabled high-performance computing holds promises in resolving the big data challenge. Third, the evaluation of geocoding accuracy, particularly for handling massive data, remains as a challenge. Spatial or spatiotemporal statistics may provide support for evaluating the robustness of the geocoding process.

References

Armstrong, M. P., G. Rushton, and D. L. Zimmerman. 1999. Geographically masking health data to preserve confidentiality. *Statistics in Medicine* 18 (5): 497–525.

Bichler, G., and S. Balchak. 2007. Address matching bias: Ignorance is not bliss. *Policing: An International Journal of Police Strategies & Management* 30 (1): 32–60.

Bonner, M. R., D. Han, J. Nie, P. Rogerson, J. E. Vena, and J. L. Freudenheim. 2003. Positional accuracy of geocoded addresses in epidemiologic research. *Epidemiology* 14 (4): 408–412.

Boscoe, F. P., M. H. Ward, and P. Reynolds. 2004. Current practices in spatial analysis of cancer data: Data characteristics and data sources for geographic studies of cancer. *International Journal of Health Geographics* 3 (1): 28.

Casas, I., E. Delmelle, and A. Varela. 2010. A space-time approach to diffusion of health service provision information. *International Regional Science Review* 33 (2): 134–156.

Chainey, S., and J. Ratcliffe. 2013. In *GIS and Crime Mapping*. London, UK: John Wiley & Sons, pp. 1–448.

Chakraborty, J., and P. A. Zandbergen. 2007. Children at risk: Measuring racial/ethnic disparities in potential exposure to air pollution at school and home. *Journal of Epidemiology and Community Health* 61 (12): 1074–1079.

Chow, T. E., N. Dede-Bamfo, and K. R. Dahal. 2016. Geographic disparity of positional errors and matching rate of residential addresses among geocoding solutions. *Annals of GIS* 22 (1): 29–42.

Cui, Y. 2013. A systematic approach to evaluate and validate the spatial accuracy of farmers market locations using multi-geocoding services. *Applied Geography* 41: 87–95.

Curtis, A. J., J. W. Mills, and M. Leitner. 2006. Spatial confidentiality and GIS: Re-engineering mortality locations from published maps about Hurricane Katrina. *International Journal of Health Geographics* 5 (1): 44.

Davis, C. A., and F. T. Fonseca. 2007. Assessing the certainty of locations produced by an address geocoding system. *Geoinformatica* 11 (1): 103–129.

Delmelle, E., C. Dony, I. Casas, M. Jia, and W. Tang. 2014. Visualizing the impact of space-time uncertainties on dengue fever patterns. *International Journal of Geographical Information Science* 28 (5): 1107–1127.

Delmelle, E. M., C. H. Cassell, C. Dony, E. Radcliff, J. P. Tanner, C. Siffel, and R. S. Kirby. 2013. Modeling travel impedance to medical care for children with birth defects using geographic information systems. *Birth Defects Research Part A: Clinical and Molecular Teratology* 97 (10): 673–684.

Delmelle, E. M., H. Zhu, W. Tang, and I. Casas. 2014. A web-based geospatial toolkit for the monitoring of dengue fever. *Applied Geography* 52: 144–152.

DeLuca, P., and P. S. Kanaroglou. 2015. An assessment of online geocoding services for health research in a mid-sized Canadian city. In *Spatial Analysis in Health Geography*, edited by P. Kanaroglou, E. Delmelle and A. Paez, New York, NY: Ashgate Publishing, pp. 31–46.

Derekenaris, G., J. Garofalakis, C. Makris, J. Prentzas, S. Sioutas, and A. Tsakalidis. 2001. Integrating GIS, GPS and GSM technologies for the effective management of ambulances. *Computers, Environment and Urban Systems* 25 (3): 267–278.

Dueker, K. J. 1974. Urban geocoding. *Annals of the Association of American Geographers* 64 (2): 318–325.

Duncan, D. T., M. C. Castro, J. C. Blossom, G. G. Bennett, and S. L. Gortmaker. 2011. Evaluation of the positional difference between two common geocoding methods. *Geospatial Health* 5(2): 265–273.

Gittler, J. 2007. Cancer registry data and geocoding. In *Geocoding Health Data: The Use of Geographic Codes in Cancer Prevention and Control, Research and Practice*, edited by G. Rushton et al., Boca Raton, FL: CRC Press, pp. 195–223.

Goldberg, D. 2008. Privacy and confidentiality. In *A Geocoding Best Practices Guide*, edited by R. Borchers et al., Springfield, IL: North American Association of Central Cancer Registries, Inc. pp. 183–187.

Goldberg, D. W. 2011. Advances in geocoding research and practice. *Transactions in GIS* 15 (6): 727–733.

Goldberg, D. W., and M. G. Cockburn. 2010. Improving geocode accuracy with candidate selection criteria. *Transactions in GIS* 14 (s1): 149–176.

Goldberg, D. W., J. P. Wilson, and C. A. Knoblock. 2007. From text to geographic coordinates: The current state of geocoding. *URISA-WASHINGTON DC-* 19 (1): 33.

Griffin, D. H., J. M. Pausche, E. B. Rivers, A. Tillman, and J. Treat. 1990. Improving the coverage of addresses in the 1990 census: Preliminary results. In *Proceedings of the American Statistical Association Survey Research Methods Section*, Anaheim, CA, 541–546.

Ha, S., H. Hu, L. Mao, D. Roussos-Ross, J. Roth, and X. Xu. 2016. Potential selection bias associated with using geocoded birth records for epidemiologic research. *Annals of Epidemiology* 26 (3): 204–211.

Hampton, K. H., M. K. Fitch, W. B. Allshouse, I. A. Doherty, D. C. Gesink, P. A. Leone, M. L. Serre, and W. C. Miller. 2010. Mapping health data: Improved privacy protection with donut method geomasking. *American Journal of Epidemiology*, 172 (9): 1062–1069.

Harada, Y., and T. Shimada. 2006. Examining the impact of the precision of address geocoding on estimated density of crime locations. *Computers & Geosciences* 32 (8): 1096–1107.

Jacquez, G. M., and J. A. Jacquez. 1999. Disease clustering for uncertain locations. *Disease Mapping and Risk Assessment for Public Health Decision Making*, edited by A. Lawson et al., London, UK: Jonh Wiley & Sons, pp. 151–168.

Jung, H., K. Lee, and W. Chun. 2006. Integration of GIS, GPS, and optimization technologies for the effective control of parcel delivery service. *Computers & Industrial Engineering* 51 (1): 154–162.

Karimi, H. A., M. H. Sharker, and D. Roongpiboonsopit. 2011. Geocoding recommender: An algorithm to recommend optimal online geocoding services for applications. *Transactions in GIS* 15 (6): 869–886.

Kirby, R. S., E. Delmelle, and J. M. Eberth. 2017. Advances in spatial epidemiology and geographic information systems. *Annals of Epidemiology* 27 (1): 1–9.

Kounadi, O., T. J. Lampoltshammer, M. Leitner, and T. Heistracher. 2013. Accuracy and privacy aspects in free online reverse geocoding services. *Cartography and Geographic Information Science* 40 (2): 140–153.

Krieger, N., J. T. Chen, P. D. Waterman, M.-J. Soobader, S. Subramanian, and R. Carson. 2002. Geocoding and monitoring of US socioeconomic inequalities in mortality and cancer incidence: Does the choice of area-based measure and geographic level matter? The Public Health Disparities Geocoding Project. *American Journal of Epidemiology* 156 (5): 471–482.

Krieger, N., P. Waterman, J. T. Chen, M.-J. Soobader, S. Subramanian, and R. Carson. 2002. Zip code caveat: Bias due to spatiotemporal mismatches between zip codes and us census–defined geographic areas—the public health disparities geocoding project. *American Journal of Public Health* 92 (7): 1100–1102.

Law, D. G., M. L. Serre, G. Christakos, P. A. Leone, and W. C. Miller. 2004. Spatial analysis and mapping of sexually transmitted diseases to optimise intervention and prevention strategies. *Sexually Transmitted Infections* 80 (4): 294–299.

Levine, N., and K. E. Kim. 1998. The location of motor vehicle crashes in Honolulu: A methodology for geocoding intersections. *Computers, Environment and Urban Systems* 22 (6): 557–576.

Luo, W., and Y. Qi. 2009. An enhanced two-step floating catchment area (E2SFCA) method for measuring spatial accessibility to primary care physicians. *Health & Place* 15 (4): 1100–1107.

Mak, S., D. T. Duncan, M. C. Castro, and J. C. Blossom. 2012. Geocoding-protected health information using online services may compromise patient privacy-Comments on "Evaluation of the positional difference between two common geocoding methods" by Duncan et al.—Response. *Geospatial Health* 6 (2): 157–159.

Mazumdar, S., G. Rushton, B. J. Smith, D. L. Zimmerman, and K. J. Donham. 2008. Geocoding accuracy and the recovery of relationships between environmental exposures and health. *International Journal of Health Geographics* 7 (1): 13.

McElroy, J. A., P. L. Remington, A. Trentham-Dietz, S. A. Robert, and P. A. Newcomb. 2003. Geocoding addresses from a large population-based study: Lessons learned. *Epidemiology* 14 (4): 399–407.

Morrill, R. L., and F. R. Pitts. 1967. Marriage, migration, and the mean information field: A study in uniqueness and generality. *Annals of the Association of American Geographers* 57 (2): 401–422.

Murray, A. T., T. H. Grubesic, R. Wei, and E. A. Mack. 2011. A hybrid geocoding methodology for spatio-temporal data. *Transactions in GIS* 15 (6): 795–809.

O'Reagan, R. T. and Saalfeld, A. 1987. Geocoding theory and practice at the bureau of the census. *Statistical Research Report Census/SRD/RR-87/29*, US. Census Bureau, Washington, DC. pp. 1–14.

Park, S. H., J. M. Bigham, S.-Y. Kho, S. Kang, and D.-K. Kim. 2011. Geocoding vehicle collisions on Korean expressways based on postmile referencing. *KSCE Journal of Civil Engineering* 15 (8): 1435–1441.

Qin, X., S. Parker, Y. Liu, A. J. Graettinger, and S. Forde. 2013. Intelligent geocoding system to locate traffic crashes. *Accident Analysis & Prevention* 50: 1034–1041.

Ratcliffe, J. 2002. Intelligence-led policing and the problems of turning rhetoric into practice. *Policing & Society* 12 (1): 53–66.

Rehkopf, D. H., L. T. Haughton, J. T. Chen, P. D. Waterman, S. Subramanian, and N. Krieger. 2006. Monitoring socioeconomic disparities in death: Comparing individual-level education and area-based socioeconomic measures. *American Journal of Public Health* 96 (12): 2135–2138.

Roongpiboonsopit, D., and H. A. Karimi. 2010a. Comparative evaluation and analysis of online geocoding services. *International Journal of Geographical Information Science* 24 (7): 1081–1100.

Roongpiboonsopit, D., and H. A. Karimi. 2010b. Quality assessment of online street and rooftop geocoding services. *Cartography and Geographic Information Science* 37 (4): 301–318.

Rushton, G., M. P. Armstrong, J. Gittler, B. R. Greene, C. E. Pavlik, M. M. West, and D. L. Zimmerman. 2006. Geocoding in cancer research: A review. *American Journal of Preventive Medicine* 30 (2): S16–S24.

Tellman, N., E. R. Litt, C. Knapp, A. Eagan, J. Cheng, and J. Lewis Jr. 2010. The effects of the Health Insurance Portability and Accountability Act privacy rule on influenza research using geographical information systems. *Geospatial Health* 5 (1): 3–9.

Werner, P. 1974. National geocoding. *Annals of the Association of American Geographers* 64 (2): 310–317.

Whitsel, E. A., K. M. Rose, J. L. Wood, A. C. Henley, D. Liao, and G. Heiss. 2004. Accuracy and repeatability of commercial geocoding. *American Journal of Epidemiology* 160 (10): 1023–1029.

Wu, J., T. H. Funk, F. W. Lurmann, and A. M. Winer. 2005. Improving spatial accuracy of roadway networks and geocoded addresses. *Transactions in GIS* 9 (4): 585–601.

Yang, D.-H., L. M. Bilaver, O. Hayes, and R. Goerge. 2004. Improving geocoding practices: Evaluation of geocoding tools. *Journal of Medical Systems* 28 (4): 361–370.

Zandbergen, P. A. 2007. Influence of geocoding quality on environmental exposure assessment of children living near high traffic roads. *BMC Public Health* 7 (1): 37.

Zandbergen, P. A. 2008. A comparison of address point, parcel and street geocoding techniques. *Computers, Environment and Urban Systems* 32 (3): 214–232.

Zandbergen, P. A. 2011. Influence of street reference data on geocoding quality. *Geocarto International* 26 (1): 35–47.

Zimmerman, D. L. 2008. Estimating the intensity of a spatial point process from locations coarsened by incomplete geocoding. *Biometrics* 64 (1): 262–270.

Zimmerman, D. L., and J. Li. 2010. The effects of local street network characteristics on the positional accuracy of automated geocoding for geographic health studies. *International Journal of Health Geographics* 9 (1): 10.

Zimmerman, D. L., J. Li, and X. Fang. 2010. Spatial autocorrelation among automated geocoding errors and its effects on testing for disease clustering. *Statistics in Medicine* 29 (9): 1025–1036.

3

Deep Learning with Satellite Images and Volunteered Geographic Information

Jiaoyan Chen and Alexander Zipf

CONTENTS

3.1 Introduction

Satellite images are images of the whole or part of the earth taken by physical sensors on satellites. Machine learning, especially deep learning (DL), has been widely applied in pattern recognition with satellite images. Many real-word applications like ground object detection, land use monitoring, and sense understanding are modeled as satellite image classification problems and successfully solved with DL algorithms. One famous application is the population distribution prediction using high-resolution satellite images and deep Convolutional Neural Networks (CNNs) by Facebook's Connectivity Lab (Gros and Tiecke 2016).

Satellite image classification refers to the task of extracting information classes (e.g., what object does an image contain?) from the satellite image. For this problem, machine learning based solutions extract input–output data pairs and then train a prediction model. Traditional machine learning models like Support Vector Machine (SVM) require the data scientists to first extract different kinds of features (e.g., texture and color) as input. This is

called feature engineering, which is labor intensive, while DL models like
AutoEncoder (AE) can automatically learn useful features from big training
data and then stack different classification layers (e.g., Logistic Regression
classifier) to represent the complex relations between data variables (Hinton
2007; Bengio, Courville, and Vincent 2013). In short, traditional machine
learning models require extraction of rich features with strong background
knowledge, while DL methods depend on big data for automatic feature representation and deep network learning.

Volunteered Geographic Information (VGI) includes a suite of tools to
create, assemble, and disseminate geographic data provided voluntarily
by individuals (Goodchild 2007). VGI systems like OpenStreetMap (OSM)
(Haklay and Weber 2008) and WikiMapia (Koriakine and Saveliev 2008) contain massive volunteered labeled ground objects with rich information like
contour, key/value tag, changeset (i.e., time-series changing records of an
object), relation (i.e., an organized list of objects for representing logic or geographic objects like bus route), etc., all of which are human knowledge about
geography. By July 2016, OSM had over 2.8 million accumulated registered
users, 3.25 billion accumulated nodes, and 250 million accumulated ways
(Allison and Jon 2016). Meanwhile, the current location-based services like
Foursquare and Strava make it possible to collect another kind of VGI data,
namely citizens' or even devices' location records, which make it possible
to discover some spatial patterns (e.g., land use type) and enrich the digital earth data (Craglia, Ostermann, and Spinsanti 2012). All these VGI data
provide an easier way to extract large labeled sample sets for training deep
models. For example, Mnih and Hinton extracted vector data from OSM
and aligned them with satellite images to train deep neural networks which
are further applied to automatically detect roads and buildings on satellite
images (Mnih and Hinton 2010; 2012).

On the other hand, using VGI together with satellite images for DL brings
new technical challenges. One challenge comes from the noise of VGI data.
Since VGI data are mostly contributed by volunteers instead of domain
experts, there are sometimes data quality problems like position inaccuracy
and classification ambiguity (Ali et al. 2014; Fan et al. 2014). Therefore, the
training samples extracted from VGI data may contain more noise than typical satellite image benchmarks like the University of California (UC) Merced
Land Use Dataset (Yang and Newsam 2010). (Mnih and Hinton 2012) classified the noise of the training samples extracted from OSM for building
detection into registration error and missing error. The former indicates the
cases that the building polygon contours on the map do not totally match the
buildings on the satellite images, while the latter represents the cases that the
buildings appearing on the satellite images are missing on the map.

Another challenge lies in *domain adaptation* (Ben-David et al. 2010), which
is known as a machine learning problem that the data for training model
differ from the data for prediction. According to the crowdsourcing study
by Quattrone et al. (Quattrone, Capra, and De Meo 2015), there is significant

geographic bias in the OSM contents. For example, most buildings in urban areas in developed countries are labeled, while the cottages in rural areas in Africa are mostly not labeled. This means that the training data from such VGI platforms will differ from the testing data that will be predicted. Except for geographic bias, the problem of domain adaptation also exists in transferring features learned from VGI data and satellite images to some other cross-domain or cross-region applications like poverty mapping and traffic prediction (Xie et al. 2016; Zhao and Kusumaputri 2016).

This chapter introduces the current DL studies with satellite images and VGI data. It first presents the classic work in satellite image classification using DL algorithms as well as some typical satellite image classification benchmarks (cf. Section 3.2), and then introduces the state of the art in utilizing VGI data as supervision knowledge for training deep neural networks, where the solutions to the technical challenges are highlighted (cf. Section 3.3). This chapter then presents some real-world applications in domains like urban computing and humanitarian mapping (cf. Section 3.4). Finally, this chapter gives the conclusion and discusses the future research directions (cf. Section 3.6).

3.2 Satellite Image Classification with Deep Learning

3.2.1 Algorithms

The general framework of DL-based methods for satellite image classification can usually be described as three components: prepared input data, deep networks, and expected output, as shown in Figure 3.1. The input data include

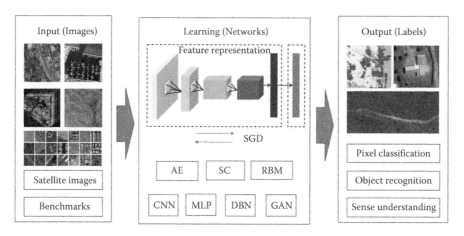

FIGURE 3.1
General framework of satellite image classification using deep learning.

different kinds of satellite images (e.g., RGB image, hyperspectral image, synthetic aperture radar image) as well as standard image datasets (i.e., benchmarks used for training), while the output is the manually defined label of the pixels or the image tiles (e.g., the land cover type of a pixel and the object type in the image tile). The deep networks are stacked by multiple nonlinear neural layers, where each intermediate layer encodes input data to feature or encode low-level features to high-level features, while the output layer predicts the probability of each label with classic classification algorithms like Logistic Regression. The parameters of each intermediate layer are usually first learned by DL models like an AE or a sparse coding algorithm with a large set of unlabeled training images, while the overall parameters are fine-tuned by algorithms like Stochastic Gradient Descent (SGD) with the supervision of labeled training images (Hinton 2007; Bengio, Courville, and Vincent 2013).

A recent literature survey about satellite image classification with DL can be found in L. Zhang, Zhang, and Kumar 2016, where classic deep networks like CNNs, Multiplayer Perceptions (MLPs), and Deep Belief Networks (DBNs) as well as feature learning algorithms like AEs, Restricted Boltzmann Machines (RBMs), and Sparse Coding (SC) are introduced and the studies are classified into four kinds according to the purpose, namely image preprocessing, pixel-based classification, target recognition, and scene understanding. Except for the above deep networks mentioned in (L. Zhang, Zhang, and Kumar 2016), we supplement another unsupervised deep model named Generative Adversarial Networks (GANs) which have been widely explored in the past two years (Goodfellow et al. 2014). A recent study by Lin (Lin 2016) showed that GANs can obtain better results than the state of the art in UC Merced Land Use Dataset and Brazilian Coffee Scenes dataset.

3.2.2 Benchmarks

The classic satellite image classification studies introduced above rely on standard datasets to train the prediction models. These studies either manually label training images with specific tools by themselves (e.g., the SAT-4 and SAT-6 datasets used in the experiments of DeepSat (Basu et al. 2015)) or adopt existing benchmarks which are often carefully made with the help of domain experts. Table 3.1 gives an overview of four widely used benchmarks for satellite image classification, where Brazilian Coffee Scenes Dataset and UCI Statlog Landsat Satellite Dataset are about land use types in rural areas, UC Merced Land Use Dataset contains different ground objects in both urban and rural areas, and SpaceNet Dataset is for buildings in urban areas. On the one hand, many kinds of ground objects such as crossroad and cottage are not included in these benchmarks, and the spatial coverage is limited to some specific areas instead of the whole world. On the other hand, the number of images is not large, which will restrict the application of the deep learning algorithms. Actually, creating a complete benchmark that can be applied in different applications will require much expert labor and is almost impossible.

TABLE 3.1

Benchmarks for Satellite Image Classification

Name	Size	Classes	Description
UC Merced Land Use Dataset (Yang and Newsam 2010)	2100 samples (100 images for each class)	21 classes: agricultural, air plane, baseball diamond, beach, buildings, chaparral, dense residential, forest, freeway, golf course, harbor, intersection, medium residential, mobile homepark, overpass, parking lot, river, runway, sparse residential, storage tanks, tennis court	Source: USGS National Map Urban Area Imagery Pixel Resolution: 0.3 m Sample input: an image with the size of 256 × 256 pixels Sample output: the class associated with the image
Brazilian Coffee Scenes Dataset (Penatti, Nogueira, and dos Santos 2015)	2876 samples (1438 image tiles for each class)	2 classes: coffee crops and none-coffee crops	Source: A composition of scenes taken by a SPOT sensor in 2005 over four counties in the State of Minas Gerais, Brazil. Sample input: an image tile with the size of 56 × 56 pixels Sample output: the class associated with the image tile
UCI Statlog Landsat Satellite Dataset (Lichman 2013)	6435 samples	7 classes: red soil, cotton crop, gray soil, damp gray soil, soil with vegetation stubble, mixture class, very damp gray soil	Source: Landsat Imagery Pixel resolution: about 80m Sample input: the multispectral values of pixels in 3 × 3 neighborhoods Sample output: the class associated with the central pixel
SpaceNet Dataset (Allison and Jon 2016)	220,594 building footprints	Building contours (polygons)	Source: 1900 square kilometers imagery (8-band multispectral data) collected by DigitalGlobe's World View-2 commercial satellite in Rio de Janeiro, Brazil Pixel resolution: 0.5 m Sample input: a 3-band or 8-band GeoTIFF file Sample output: vector data about building contours in the GeoJSON format

3.3 VGI for Deep Learning

3.3.1 VGI Data Quality and Noise

Since the data on VGI sites are mostly contributed by common volunteers, the data quality problem has attracted a lot of concerns. Haklay (Haklay 2010) analyzed the data quality of OSM in London, where, for example, he found that the motorway objects of OSM approximately had an 80% overlap in comparison with Ordnance Survey datasets. Girres and Touya (Girres and Touya 2010) extended the above OSM data quality analysis to France with more quality elements, where BD TOPO® data were used as the reference. On the basis of the measurements over OSM data (Girres and Touya 2010; Haklay 2010), we list the main aspects of the data quality problem:

- *Geometric inaccuracy*: the position of the mapped objects is not completely right.
- *Attribute inaccuracy*: objects are not fully informed by some important tag or are wrongly informed.
- *Semantic inaccuracy*: the nature or function of the mapped objects is not correctly labeled.
- *Incompleteness*: not all the ground objects are mapped by the volunteers.
- *Logical inconsistency*: the spatial relation of two mapped objects sometimes does not satisfy the common sense in reality.
- *Temporal inaccuracy*: the changes of the real objects are not timely mapped.
- *Lineage missing*: the source information of the mapped object is not attached.

When VGI data are utilized as the input of machine learning algorithms, these data quality issues lead to three kinds of noise among the extracted training samples, as shown in Table 3.2. The omission noise and registration

TABLE 3.2

Noise of Training Samples Extracted from VGI Data

Noise Type	Definition	Data Quality Sources
Omission noise	The training sample set is insufficient to estimate the real data distribution	Incompleteness, Temporal inaccuracy, Attribute inaccuracy
Registration noise	The input of the training sample is biased	Geometric inaccuracy, Temporal inaccuracy, Logic inconsistency
Semantic noise	The training sample is wrongly labeled	Semantic inaccuracy, Attribute inaccuracy

noise are proposed by Mnih and Hinton (Mnih and Hinton 2012) in their study of extracting labels from OSM for deep learning, while the semantic noise is proposed by the authors for the cases of extracting wrong samples for a class.

VGI researchers have proposed quite a few solutions that can help improve the quality of VGI data, some of which can be utilized to pre-process the VGI data before they are applied in deep learning applications. First, many crowd sourcing sites record the volunteers' behavior in specific data structure such as the Changeset of OSM. Such additional information makes it possible to extract the mapped objects at the specific time (Allison and Jon 2016) when the satellite images are taken, which can mainly help reduce the registration noise caused by temporal inaccuracy.

Second, quite a few VGI data quality evaluation and data quality improve-ment methods have been proposed. In the study by Ali et al. (Ali et al. 2014), a machine learning-based method is proposed to track OSM classification plausibility such as the wrong semantic labels to ambiguous areas, which can reduce the semantic noise of the training sample when applied in pre-processing the VGI data. More such intrinsic methods for VGI data quality can be found in the study by Barron et al. (Barron, Neis, and Zipf 2014).

Third, linking multiple VGI data provides another way to improve the data quality, especially for incompleteness. For example, GPS traces from citizen sensors, for example, taxi cars, provide an alternative way of road map gen-eration (L. Zhang, Thiemann, and Sester 2010). It can supplement the road information of OSM, thus reducing the registration noise of the training sample set for road detection.

3.3.2 Learning Algorithms for the Noise

Machine learning researchers have proposed some more robust learning algorithms with the noise considered for more robust prediction models. We first formally describe and model the satellite image classification problem following the road detection study by Mnih and Hinton (Mnih and Hinton 2010). The problem is simplified as a binary classification of each pixel, that is, positive for the object of interest and negative for the NOT interested. A satellite image is denoted by $\mathbf{S}$, while a map of equal size is denoted by $\tilde{\mathbf{M}}$, where $\tilde{\mathbf{M}}_{i,j} = 1$ if the pixel at location (i, j) is positive and $\tilde{\mathbf{M}}_{i,j} = 0$ otherwise. Two vectors s and $\tilde{m}$ are used to denote the image patch $n(\mathbf{S}_{i,j}, w_s)$ with the center location of (i, j) and the size of $w_s \times w_s$ and the map patch $n(\mathbf{M}_{i,j}, w_m)$. The conditional probability distribution of the map pixels can be described as:

$$p(\tilde{\mathbf{m}} \,|\, \mathbf{s}) = \prod_{i=1}^{w_m^2} p(\tilde{m}_i \,|\, \mathbf{s}) \tag{3.1}$$

where each $p(\tilde{m}_i \,|\, \mathbf{s})$ is assumed to follow the Bernoulli distribution and its mean value is determined by the ith output node when the distribution is modeled by a neural network which can be stacked with different layers like the fully connected layer and the subsampling layer. Learning can be implemented by optimizing an object function, for example, negative log likehood optimization function with the following cross entropy format over the training samples with optimization algorithms like the batched stochastic gradient descent:

$$\sum_{i=1}^{w_m^2} (\tilde{m}_i \ln \hat{m}_i + (1 - \tilde{m}_i)\ln(1 - \hat{m}_i)) \tag{3.2}$$

where $\hat{m}_i$ represents the pixel of the predicted map batch.

The omission noise is dealt with in (Mnih and Hinton 2012) with a model named the asymmetric Bernoulli noise (ABN) model which rewrites Equation 3.1 as:

$$p(\tilde{\mathbf{m}} \mid \mathbf{s}) = p(\tilde{\mathbf{m}} \mid \mathbf{m})p(\mathbf{m} \mid \mathbf{s}) = \prod_{i=1}^{w_m^2} p(\tilde{m}_i \mid m_i)p(m_i \mid \mathbf{s}) \tag{3.3}$$

where $\mathbf{m}$ and $\tilde{\mathbf{m}}$ are respectively known as the truth map batch and the observed map batch. ABN assumes that conditioned on $\mathbf{m}$, all the components of $\tilde{\mathbf{m}}$ are independent and $\tilde{m}_i$ is assumed to be independent of m_j for any $j \neq i$, and the noise distribution $p(\tilde{m}_i \mid m_i)$ is assumed to be the same for all pixels i. By setting $\theta_0 \ll \theta_1$, where $\theta_0 = p(\tilde{m}_i = 1 \mid m_i = 0)$ and $\theta_1 = p(\tilde{m}_i = 0 \mid m_i = 1)$, the omission noise can be managed in learning as this can reduce the false negative rate caused by the omission noise.

The above ABN model is extended in (Mnih and Hinton 2012) as a translational ABN (TABN) model to deal with the registration noise. $p(\tilde{\mathbf{m}} \mid \mathbf{m})$ in Equation 3.3 is extended with the translational noise considered as $p(\tilde{\mathbf{m}} \mid \mathbf{m}, t) = p(\tilde{\mathbf{m}} \mid \text{Crop}(\mathbf{m}, t))$, where $\text{Crop}(\mathbf{m}, t)$ function selects $w_m' \times w_m'$ subpatch from $w_m \times w_m$ patch $\mathbf{m}$ according to the translation variable t, as shown in Figure 3.2. Finally, the conditional distribution, that is, Equation 3.3 is extended as:

$$p(\tilde{\mathbf{m}} \mid \mathbf{s}) = \sum_{t=0}^{T} p(t) \sum_{\mathbf{m}} p(\tilde{\mathbf{m}} \mid \mathbf{m}, t)p(\mathbf{m} \mid \mathbf{s}) \tag{3.4}$$

where parameters in $p(t)$ and $p(\tilde{\mathbf{m}} \mid \mathbf{m}, t)$ are set using a validation sample set, while the parameters in $p(\mathbf{m} \mid \mathbf{s})$ are learned by mining the negative log likehood object function with the EM-algorithm over the training sample set. In the perspective of geometry, the translation function Crop randomly selects different subparts of the true map generalized from the satellite image for further generalizing the observed map.

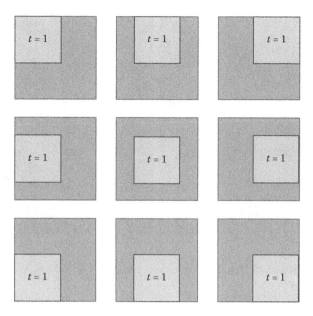

FIGURE 3.2
Demonstration of the function of Crop used in the TABN model. The outside patches represent m while the inner patches represent the cropped patch Crop (m,t). (From Mnih, V. and G. Hinton. 2012. Learning to label aerial images from noisy data. In *Proceedings of the 29th International Conference on Machine Learning—ICML'12*, Edinburgh, UK, 567–574.)

Yuan (Yuan 2016) introduced a signed distance function to the loss function for building detection with convolutional networks, which we think is helpful for dealing with the registration noise. In detail, the loss item of each pixel m_i of the training data in Equation 3.2 is assigned a signed distance d_i to the boundary: $d_i = 0$ if the ith pixel is on the boundary of the building footprint, $d_i > 0$ if it is inside the building footprint, and $d_i < 0$ if it is outside the building footprint, as shown in Figure 3.3. According to (Yuan 2016), the signed distance allows the deep network to learn more spatial layout information, for example, the point inside the footprint is more important than the one close to the boundary. This means the registration noise which usually lies on the boundary will have less impact on the model than the samples that are more likely to be true, that is, the pixels inside.

3.3.3 Spatial and Semantic Domain Adaptation

On VGI sites, crowdsourcing is often spatially or semantically biased (Quattrone, Capra, and De Meo 2015). For example, OSM volunteers contribute more to the urban areas than to the rural areas, and more to the public points of interest (POIs), for example, shopping malls, than to the residential areas. When VGI data are applied in deep learning applications, such bias phenomena will lead to the problem of *domain adaptation*

FIGURE 3.3
(a) Training image and (b) corresponding signed distance label. (Allison, G. and B. Jon. 2016. Exploring the SpaceNet Dataset Using DIGITS. https://devblogs.nvidia.com/parallelforall/exploring-spacenet-dataset-using-digits/.)

(Ben-David et al. 2010) which aims at learning a model from a source data distribution that can perform well on a related but different target data distribution. In a dynamic or streaming context, the data difference between the training source and the target is also known as the problem of *concept drift* (Tsymbal 2004).

There are quite a few typical algorithms for domain adaptation, such as reweighting the source samples, iteratively labeling the target examples with new models, and searching for a common data distribution space between the source and the target. These methods improve the learning algorithm itself without taking the properties of VGI data into consideration and actually are applicable for any data. In this chapter, we introduce two other techniques, namely active sampling and feature transferring for deep learning over the biased VGI data.

Active sampling is a strategy for active learning which aims at building efficient training sample sets by querying the user or some other data sources. It has been studied for years in satellite image classification with deep neural networks as a cost considered strategy for manual labeling of the training samples, as shown in the survey by Tuia, et al. (Tuia et al. 2011). We introduce a VGI-based active deep learning framework used in our study named DeepVGI (Chen and Zipf 2017) which aims at learning from Bing satellite images (RGB, level 18, 256*256), MapSwipe data (volunteer label of the Bing satellite image), and OSM data for humanitarian mapping, which is house and building detection in rural areas in Africa. Instead of classifying each pixel modeled in Section 3.3.2, DeepVGI has the same task as the MapSwipe volunteers, namely directly classifying the image into positive (i.e., containing the target object) and negative (i.e, NOT

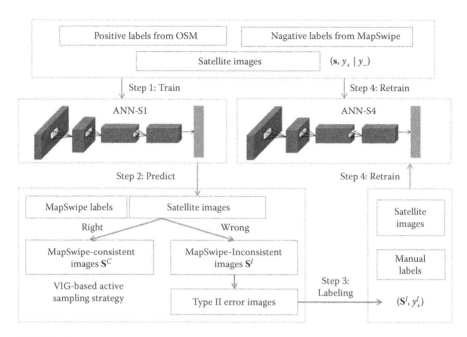

FIGURE 3.4
Workflow of DeepVGI with a VGI-based active sampling strategy.

containing any target objects). As shown in Figure 3.4, DeepVGI first trains a deep network with positive samples from OSM and negative samples from MapSwipe, and then uses the network to predict the training images. The images whose predicted label is the sample as the MapSwipe volunteers are called the MapSwipe-Consistent $\mathbf{S}^C$, while the others are called the MapSwipe-Inconsistent $\mathbf{S}^I$. DeepVGI then asks the experts to label a number of images in $\mathbf{S}^I$, and further adds these new labeled images to the original samples for retraining a new deep network. The number of images for relabeling by the experts is limited to reduce the cost. Such an active learning strategy reduces the omission noise of OSM by experts' labeling and MapSwipe labels. In the experiment of building detection in Africa, the retrained network (ANN-S4) is well generalized from the buildings downtown to the cottages in rural areas, thus achieving a higher overall prediction accuracy than the first network (ANN-S1) trained without any actively sampled images.

Transfer learning, which transfers features learned in one domain to another domain, provides another solution for domain adaptation (Pan and Yang 2010). In satellite image classification, there have been several transfer learning studies for domain adaptation. Jun and Ghosh proposed to transfer the knowledge learned from one region to another (Jun and Ghosh 2008), while Demir et al. (Demir, Bovolo, and Bruzzone 2013) proposed to transfer the features learned from one time to another. Yuan et al. (Yuan and Cheriyadat 2014) transferred the low-level features, for example, line segment learned from satellite images to predict the number of buildings. Researchers from Facebook's Connectivity Lab found that, with minor modifications, the Convolutional Neural Networks trained on normal photos can efficiently detect whether a satellite image contains a footprint, which means the deep features learned from normal photos can be transferred to deep learning tasks over satellite images. However, the three studies introduced above did not utilize VGI data as the supervison knowledge. As far as we know, there are currently no studies that transfer learned VGI knowledge from task to task, from area to area, or from time to time in satellite image classification with deep neural networks.

3.4 Applications

By integrating VGI data and satellite images, deep learning technologies enable the machine to learn the crowdsourcing knowledge contributed by the volunteers and citizens, thus automatically and intelligently exploring our earth and society. Many novel applications can be proposed with the data and the technologies. Here, we only present some typical examples.

- *Humanitarian mapping*: Many ground objects like road, house, rivers, etc. in rural and undeveloped areas are still missing in the current maps, but they are quite important to help the people in need. We can learn ground object prediction models from the existing map data (Mnih and Hinton 2010; Mnih and Hinton 2012; Roemheld 2016; Chen and Zipf 2017), and then predict the missing object on the satellite images, to either save the volunteers' labor by recommendation or improve the quality of the data contributed by the volunteers (Ali et al. 2014).

- *Population mapping*: The spatial distribution of a population is quite important in making some decisions. For example, Facebook needs to know how many devices shoud be deployed in each area so as to connect the whole world. Buildings and roads on satellite images or POIs from OSM provide one way for global population mapping (Bakillah et al. 2014), while VGI records from citizen sensors like the mobile phones provide another solution for dynamic population estimation for a specific area (Deville et al. 2014).

- *Poverty prediction*: Similar to population mapping, with the data of buildings and roads on satellite images or OSM, the global poverty map, which constitute critical data for studying the society and economy, can be estimated (Zhao and Kusumaputri 2016). Instead of utilizing building and road data, Xie et al. (Xie et al. 2016) adopted nighttime light intensities which were predicted by the daytime satellite images with Convolutional Neural Networks as a proxy for population prediction.

- *Urban change detection*: With footprints of ground urban objects on OSM and satellite images, we are able to detect any land cover changes such as the damage caused by an earthquake or a storm in an urban area by pixel-wise classification with deep learning techniques (Yuan 2016).

- *Public health monitoring*: Public health can be measured and monitored by citizen sensors like tweets (Paul and Dredze 2011), while many causes of public health events like smog disasters and flood disasters can be observed by satellite sensors. Connecting the physical sensor data to the citizen sensor data makes it possible to analyze and monitor the public health (Chen et al. 2014). Deep neural networks enable researchers to find complex patterns between public health and big VGI data.

3.5 Conclusion and Future Work

In this chapter, we introduced some recent work in deep learning with satellite images and VGI data. We first analyzed the typical deep learning studies in satellite image classification as well as some classic benchmarks, and then focused on the problem of automatically extracting big sample sets from VGI data for the supervision of training deep networks. Two main technical challenges about sample noise and domain adaptation as well as their solutions in VGI data quality research and machine learning research were further introduced. Finally, we presented several applications where the above techniques and data can be applied.

Learning deep prediction models from VGI data and satellite images with the supervision of volunteers' knowledge is a promising field with many potential real-world applications. On the one hand, we think the field of VGI itself will be further studied, especially the aspects of data quality and data linkage. Linking VGI data can not only make up the problem of data quality, but also enrich the supervision knowledge for different prediction tasks and fine-grained prediction models. On the other hand, feature-level data integration for more accurate prediction models is an important research direction. Machine learning algorithms like transfer learning and active learning will be applied together with deep feature representation for these technical problems.

References

Ali, A. L., F. Schmid, R. Al-Salman, and T. Kauppinen. 2014. Ambiguity and plausibility: Managing classification quality in volunteered geographic information. In *Proceedings of the 22nd SIGSPATIAL International Conference on Advances in Geographic Information Systems*, Dallas, Texas, 143–152. doi: 10.1145/2666310.2666392.

Allison, G. and B. Jon. 2016. Exploring the SpaceNet Dataset Using DIGITS. https://devblogs.nvidia.com/parallelforall/exploring-spacenet-dataset-using-digits/

Bakillah, M., S. Liang, A. Mobasheri, J. J. Arsanjani, and A. Zipf. 2014. Fine-resolution population mapping using open street map points-of-interest. *International Journal of Geographical Information Science 28 (March 2015)*. Taylor & Francis: 1–24. doi: 10.1080/13658816.2014.909045.

Barron, C., P. Neis, and A. Zipf. 2014. A comprehensive framework for intrinsic openstreetmap quality analysis. *Transactions in GIS* 18(6), Wiley Online Library: 877–895.

Basu, S., S. Ganguly, S. Mukhopadhyay, R. DiBiano, M. Karki, and R. Nemani. 2015. Deepsat: A learning framework for satellite imagery. In *Proceedings of the 23rd SIGSPATIAL International Conference on Advances in Geographic Information Systems*, Seattle, Washington, 37.

Ben-David, S., J. Blitzer, K. Crammer, A. Kulesza, F. Pereira, and J. W. Vaughan. 2010. A theory of learning from different domains. *Machine Learning* 79(1–2), Springer: 151–175.

Bengio, Y., A. Courville, and P. Vincent. 2013. Representation learning: A review and new perspectives. *IEEE Transactions on Pattern Analysis and Machine Intelligence* 35(8), IEEE: 1798–1828.

Chen, J. and A. Zipf. 2017. DeepVGI: Deep learning with volunteered geographic information. In *WWW '17 Companion: Proceedings of the 26th International Conference Companion on World Wide Web*, Perth, Australia.

Chen, J., H. Chen, G. Zheng, J. Z. Pan, H. Wu, and N. Zhang. 2014. Big smog meets web science: Smog disaster analysis based on social media and device data on the web. In *Proceedings of the 23rd International Conference on World Wide Web*, Seoul, Korea, 505–510.

Craglia, M., F. Ostermann, and L. Spinsanti. 2012. Digital earth from vision to practice: Making sense of citizen-generated content. *International Journal of Digital Earth* 5(5). Taylor & Francis: 398–416.

Demir, B., F. Bovolo, and L. Bruzzone. 2013. Updating land-cover maps by classification of image time series: A novel change-detection-driven transfer learning approach. *IEEE Transactions on Geoscience and Remote Sensing* 51(1): 300–312. doi: 10.1109/TGRS.2012.2195727.

Deville, P., C. Linard, S. Martin, M. Gilbert, F. R. Stevens, A. E. Gaughan, V. D. Blondel, and A. J. Tatem. 2014. Dynamic population mapping using mobile phone data. *Proceedings of the National Academy of Sciences* 111(45), National Acad Sciences: 15888–15893.

Fan, H., A. Zipf, Q. Fu, and P. Neis. 2014. Quality assessment for building footprints data on openstreetmap. *International Journal of Geographical Information Science* 28(4), Taylor & Francis: 700–719.

Girres, J.-F. and G. Touya. 2010. Quality assessment of the French openstreetmap dataset. *Transactions in GIS* 14(4), Wiley Online Library: 435–459.

Goodchild, M. F. 2007. Citizens as sensors: The world of volunteered geography. *GeoJournal* 69(4), Springer: 211–221.

Goodfellow, I., J. Pouget-Abadie, M. Mirza, B. Xu, D. Warde-Farley, S. Ozair, A. Courville, and Y. Bengio. 2014. Generative adversarial networks. In *Advances in Neural Information Processing Systems*, Palais des Congrès de Montréal, Montréal, Canada, 2672–2680.

Gros, A. and T. Tiecke. 2016. Connecting the World with Better Maps. https://code.facebook.com/posts/1676452492623525/connecting-the-world-with-better-maps/.

Haklay, M. 2010. How good is volunteered geographical information? A comparative study of openstreetmap and ordnance survey datasets. *Environment and Planning B: Planning and Design* 37(4): 682–703. doi: 10.1068/b35097.

Haklay, M. and P. Weber. 2008. Openstreetmap: User-generated street maps. *IEEE Pervasive Computing* 7(4), IEEE: 12–18.

Hinton, G. E. 2007. Learning multiple layers of representation. *Trends in Cognitive Sciences* 11(10), Elsevier: 428–434.

Jun, G. and J. Ghosh. 2008. An efficient active learning algorithm with knowledge transfer for hyperspectral data analysis. *IGARSS 2008 - 2008 IEEE International Geoscience and Remote Sensing Symposium*, Boston, MA, I-52-I-55. doi: 10.1109/IGARSS.2008.4778790.

Koriakine, A. and E. Saveliev. 2008. WikiMapia. Online: Wikimapia.org.

Lichman, M. 2013. UCI Machine Learning Repository. http://archive.ics.uci.edu/ml.

Lin, D. Yu. 2016. Deep Unsupervised Representation Learning for Remote Sensing Images. CoRR.

Mnih, V. and G. Hinton. 2012. Learning to label aerial images from noisy data. In *Proceedings of the 29th International Conference on Machine Learning—ICML'12*, Edinburgh, UK, 567–574.

Mnih, V. and G. E. Hinton. 2010. Learning to detect roads in high-resolution aerial images. In *European Conference on Computer Vision*, Crete, Greece. Vol 6316, pp. 210–223.

Pan, S. J. and Q. Yang. 2010. A survey on transfer learning. *IEEE Transactions on Knowledge and Data Engineering* 22(10): 1345–1359. doi: 10.1109/TKDE.2009.191.

Paul, M. J. and M. Dredze. 2011. You are what you tweet: Analyzing twitter for public health. In *The 5th International AAAI Conference on Web and Social Media*, Barcelona, Spain. Vol 20, pp. 265–272.

Penatti, O. A. B., K. Nogueira, and J. A. dos Santos. 2015. Do deep features generalize from everyday objects to remote sensing and aerial scenes domains? In *Proceedings of the IEEE Conference on Computer Vision and Pattern Recognition Workshops*, Boston, MA. pp. 44–51.

Quattrone, G., L. Capra, and P. De Meo. 2015. There's no such thing as the perfect map: Quantifying bias in spatial crowd-sourcing datasets. In *Proceedings of the 18th ACM Conference on Computer Supported Cooperative Work & Social Computing—CSCW '15*, Vancouver, Canada. pp. 1021–1032. doi: 10.1145/2675133.2675235.

Roemheld, L. 2016. Humanitarian Mapping with Deep Learning.

Tsymbal, A. 2004. The problem of concept drift: Definitions and related work. *Computer Science Department, Trinity College Dublin* 4(C): 2004–2015. doi: 10.1.1.58.9085.

Tuia, D., M. Volpi, L. Copa, M. Kanevski, and J. Munoz-Mari. 2011. A survey of active learning algorithms for supervised remote sensing image classification. *IEEE Journal of Selected Topics in Signal Processing* 5(3): 606–617. doi: 10.1109/JSTSP.2011.2139193.

Xie, M., N. Jean, M. Burke, D. Lobell, and S. Ermon. 2016. Transfer learning from deep features for remote sensing and poverty mapping. In *Proceedings of the Thirtieth AAAI Conference on Artificial Intelligence*, Phoenix, Arizona. pp. 3929–3935.

Xu, F. F., B. Y. Lin, Q. Lu, and K. Q. Zhu. 2016. Cross-region traffic prediction for China on openstreetmap. In *Proceedings of the 9th ACM SIGSPATIAL International Workshop on Computational Transportation Science*, San Francisco, California. pp. 37–42.

Yang, Y. and S. Newsam. 2010. Bag-of-visual-words and spatial extensions for land-use classification. In *Proceedings of the 18th SIGSPATIAL International Conference on Advances in Geographic Information Systems*, San Jose, California. pp. 270–279.

Yuan, J. 2016. Automatic Building Extraction in Aerial Scenes Using Convolutional Networks. CoRR. http://arxiv.org/abs/1602.06564.

Yuan, J. and A. M. Cheriyadat. 2014. Learning to count buildings in diverse aerial scenes. In *Proceedings of the 22nd ACM SIGSPATIAL International Conference on Advances in Geographic Information Systems - SIGSPATIAL '14*, Dallas, Texas. pp. 271–280. doi: 10.1145/2666310.2666389.

Zhang, L., L. Zhang, and V. Kumar. 2016. Deep learning for remote sensing data. *IEEE Geoscience and Remote Sensing Magazine* 4: 22–40.

Zhang, L., F. Thiemann, and M. Sester. 2010. Integration of GPS traces with road map. In *Proceedings of the Third International Workshop on Computational Transportation Science*, San Jose, California. pp. 17–22. IWCTS '10. New York, NY: ACM. doi: 10.1145/1899441.1899447.

Zhao, L. and P. Kusumaputri. 2016. Openstreetmap road network analysis for poverty mapping.

4

Visual Analysis of Floating Car Data

Linfang Ding, Jukka M. Krisp, and Liqiu Meng

CONTENTS

4.1 Introduction

With the advances in location positioning and wireless communication technologies, collecting spatial trajectories that represent the mobility of a variety of moving objects becomes prevalent in the digital society. Floating car data (FCD) gathered from GPS-equipped moving vehicles have become increasingly available. For instance, a large number of taxis in major cities, for example, San Francisco, Shanghai, Rome, are equipped with GPS devices and send time-stamped locations of a high frequency to data centers and result in huge amounts of FCD. With the open data movement, some FCD data sets, for example, mobility traces of taxicabs in San Francisco (Michal, Natasa, and Matthias 2009), are freely available and have been widely used for a variety of research purposes and applications. The complex and large FCD contain rich information and bring new opportunities to understand urban dynamics, which are crucial for decision-making in environmental and transportation planning.

FCD have been intensively studied for a wide range of research and application purposes. A large group of research works have been focused on utilizing FCD for modeling traffic congestion and human mobility patterns (Ding, Yang, and Meng 2015; Keler, Ding, and Krisp 2016), and for uncovering driving behaviors (Liu, Andris, and Ratti 2010; Ding, Fan, and Meng 2015). Some research works have examined FCD to infer urban land uses and city structures (Liu, Gong et al. 2015) and to mine interesting locations or places (Zheng et al. 2009; Andrienko et al. 2011). Other investigations have been conducted to understand place semantics together with other data sources, like point-of-interest (POI) data (Yuan, Zheng, and Xie 2012) and social media data (Liu, Liu et al. 2015; Mazimpaka and Timpf 2015).

Basically, FCD consist of position records generated by moving vehicles. Each record can be represented by a point of the form $p = (x, y, t)$ and associated with additional fields like velocity and orientation. A series of chronologically ordered points form a spatial trajectory $(p_1, p_2, \dots, p_n)$. FCD analysis, and in general spatial trajectory analysis, involves a variety of research topics and techniques from spatial trajectory data preprocessing to trajectory pattern mining and to big trajectory data visualization. Most of the existing research on FCD uses computational only approaches which normally lack the involvement of human interaction and effective communication of the results with the human.

This chapter addresses visual analytics approaches for FCD. Visual analytics is a fast evolving discipline of analytical reasoning facilitated by interactive visual interfaces (Thomas and Cook 2005; Keim et al. 2010). In recent years, visual analytics techniques and tools have been increasingly proposed, developed, and applied to explore big geospatial movement data for understanding human mobility patterns and urban structures.

A comprehensive exploration and understanding of massive FCD requires a variety of visual analytics techniques ranging from direct depictions of original FCD to representations of their computationally derived data. Research works on FCD visual analysis can be distinguished at two abstract levels, namely, (1) point-based and (2) trajectory-based levels. The point-based view considers each discrete point, either a raw GPS entry or a derived point spatial object, as a point-based spatial object, while the trajectory-based view considers a sequence of temporally ordered points (e.g., GPS records) as a trajectory-based spatial object.

Taking the point-based view, an FCD data point can be represented and visualized, for example, by a dot. Interactive techniques, such as filtering, brushing, and linking, can be applied to explore interesting parts of the data set. For instance, the interactive visualization system HubCab plots millions of individual pick-up and drop-off points, and allows users to get insight into the taxi mobility patterns at a very fine granularity to support taxi sharing services (Santi et al. 2014). In order to explore multiple attributes of FCD data points, multivariate visualizations (Wong and Bergeron 1997) can be utilized. At the collective level, visual analysis of massive FCD incorporates data transformation and data aggregation to represent groups of objects and reduces visual cluttering. A typical procedure for visual analysis of point objects involves point clustering, space partition, spatiotemporal aggregation, and the analysis of aggregated data. Andrienko and Andrienko (2010) systematized aggregation approaches of movement data into a framework that clearly defines what kinds of exploratory tasks each approach is suitable for.

Taking the trajectory-based view, a straightforward visualization is simply connecting the adjacent trajectory segments into lines. However, drawing many such lines may lead to overplotting so that users could hardly discern any meaningful patterns. To reduce the visual clutter, techniques such as edge bundling (Holten 2006; Holten and Van Wijk 2009; Zhou et al. 2013), animations (e.g., the NYC Taxi holiday visualization system (https://taxi. imagework.com), and Space-Time-Cube (Kraak 2003) can be applied. Visual analytics approaches (e.g., visual clustering [Andrienko et al. 2009]) and systems (e.g., TripVista [Guo et al. 2011]) driven by human analytics have been developed for the interactive exploration of large collections of trajectories. At an aggregated level, trajectories can be grouped into movement flows and visualized. Andrienko and Andrienko (2011) presented a generic spatial generalization and aggregation approach for visual analysis of movement trajectories. Composite density maps (Scheepens et al. 2011), stacked 3D trajectory bands (Tominski et al. 2012), and network visualization (van den Elzen and van Wijk 2014) were proposed to explore the multivariate trajectory data. In particular, several research works (Guo 2009; Wood, Dykes, and Slingsby 2010; Boyandin et al. 2011; Guo and Zhu 2014) were specialized in mapping origin–destination flows.

In this chapter, we introduce geospatial visual analytics techniques to explore FCD points and trajectories, especially at individual levels. The techniques are demonstrated using a large amount of real-world FCD collected in Shanghai (Ding 2016).

4.2 FCD and Preprocessing

The test FCD data set is temporally ordered position records collected from about 2000 taxis within 52 days from May 10 to June 30, 2010 in Shanghai with a temporal resolution of 10 seconds, resulting in more than half a billion GPS

TABLE 4.1

Test Data Properties

Field	Example Value	Field Description
Date	20,100,517	8-digit number, yyyymmdd
Time	235,903	6-digit number, HHMMSS
Car identifier	10,003	5-digit number
Longitude	121.472038	Accurate to 6 decimal places, in degrees
Latitude	31.236135	Accurate to 6 decimal places, in degrees
Velocity	16.1	In km/h
Car status	1/0	1-occupied; 0-unoccupied

entries. Each GPS entry is associated with fields of date, time, car identifier, location, instantaneous velocity, and car status. Table 4.1 lists the fields for each GPS record along with sample values and descriptions. Figure 4.1 illustrates the raw GPS points of a taxi with the identifier 10003 on May 12, 2010.

For each day, the size of data as a CSV file is about 4.5 G. We preprocess the raw data by filtering out the errors, for instance, GPS data outside the boundary of Shanghai, time stamps not in the valid test time slot, and the attributes that are not meaningful.

4.2.1 Classification of FCD Points

A variety of different point types can be identified based on GPS attributes. In this study, we use "car status" as an illustration. On the basis of the attribute value of "car status" 1 and 0, we can easily differentiate occupancy (O) and non-occupancy (N) points. Furthermore, we derive two additional special types of points, namely pick-up (P) and drop-off (D) points, from a time series of GPS points. A pick-up point is a location where the car status

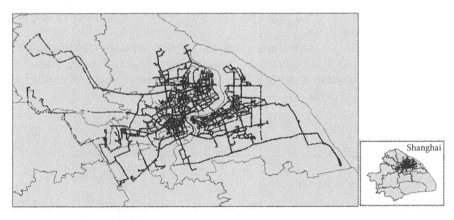

FIGURE 4.1
GPS points of the taxi with ID 10003 on May 12, 2010.

changes from non-occupancy to occupancy, while a drop-off point is a location where the status changes in the opposite way.

Owing to the huge amount of (O, N, P, D) points in the time span in the whole study area, we partition them into certain spatial (e.g., 100 × 100 m) and temporal (e.g., 1 h) chunks. For each hour in the 52 days, we compute the total numbers of (O, N, P, D) points respectively in the study area. Since the O and N points are at the same orders of magnitude, and P and D are at the same orders of magnitude, we plot in the pair of (O, N) and (P, D) separately. Figure 4.2 illustrates the temporal variation of (O, N, P, D) points. In Figure 4.2, the frequency distributions of the four types of points exhibit strong daily rhythm patterns. Besides, there is clearly a negative correlation between the distributions of (O, N) points and a positive correlation between (P, D) points. Furthermore, we can easily identify the peaks and valleys of the (O, N, P, D) distributions. For instance, the (P, D) point distribution has three peaks (at about 8–9, 12, and 18 h) and one deep valley (at about 4 h).

For each spatial chunk of 100 × 100 m, we compute the total number of (O, N, P, D) points in the 52 days. A constant interval is used to group the values into seven classes and a grayscale scheme from black to white is applied to show the distribution of the data. Figure 4.3 shows the distinct spatial distributions of the (O, N, P, D) points.

4.2.2 Extraction of Trajectories

A trajectory is a complex spatiotemporal object consisting of consecutive GPS records of an entity with several associated attributes. Similar to the

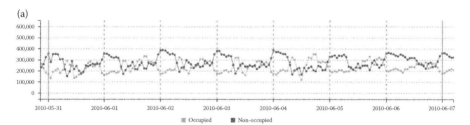

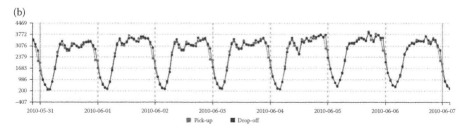

FIGURE 4.2
The hourly temporal variation of the total numbers of (O, N, P, D) in 1 week. (a) Occupancy and non-occupancy points. (b) Pick-up and drop-off points.

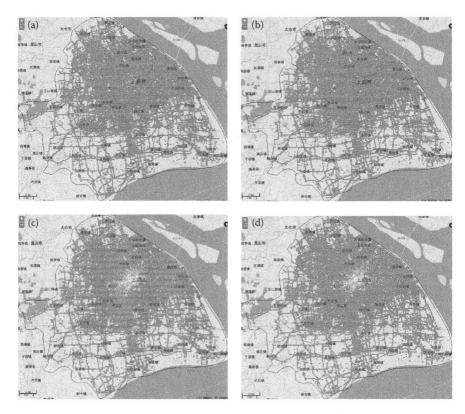

FIGURE 4.3
The frequency distribution of the total numbers of (O, N, P, D) points. (a) Occupancy. (b) Non-occupancy. (c) Pick-up. (d) Drop-off.

treatment in the point view, a variety of trajectory types can be derived from the GPS record series based on their attributes. In addition, simplified trajectories can also be derived at different abstract levels. For instance, we can easily reconstruct occupancy and non-occupancy trajectories or trips based on the "car status" values 1 and 0. By connecting the first and last points of the occupancy trajectories, we can derive their origin–destination lines. Figure 4.4 illustrates the reconstructed occupied trajectories of 100 cars on May 10, 2010, and their corresponding origin–destination lines. The trace footprints reflect the relative density of trajectories and the road network structure well. We can also perceive some very popular origins and destinations, for instance, a hotspot on the right of the screenshot, which corresponds to the Shanghai Pudong International Airport.

Besides, a number of associated trip statistics, for example, trip distance and duration, can be derived. On the basis of the occupied trip distance and the taxi fare system of Shanghai in 2010, we calculate the average daily income of each taxi, which exhibits a normal distribution (Ding, Fan, and

 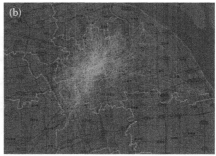

FIGURE 4.4
The reconstructed occupancy trajectories and their origin–destination lines. (a) Occupied trajectories. (b) Origin–destination lines.

Meng 2015). This chapter categorizes the 100 highest-income taxis as the top group and the 100 lowest-income taxis as the bottom group.

4.3 Point-Based Visual Analysis

As introduced in Section 4.1, there have been a variety of techniques for visual exploration and communication of FCD points at different aggregation levels. In this section, we propose a pie radar glyph for multivariate point-based visualization and a salience-based method for the dominant variate visualization.

4.3.1 Pie Radar Glyph

Glyph-based visualization methods like star/radar plot are widely used to represent multivariate data. In this work, we propose a pie radar glyph to represent the multivariate FCD points. The glyph comprises four filled sectors assigned with orange, green, red, and blue anticlockwise to represent the variates of (O, N, P, D), respectively. The radius of each sector is proportional to the value of the mapping variate, which is calculated for the total number of points within a spatial partition of 100×100 m and the temporal slot of 1 hour. Figure 4.5 depicts the mapping of a pie radar glyph from the co-located four variate data.

To avoid cluttering effects and enhance the visual appearance, we first properly order glyphs by assigning a larger z-index value to small glyphs so that they are placed on top of larger ones rather than being occluded. Then we render the glyphs using semi-transparency to make sure that the large glyphs underneath are visible. The radius of the glyph sectors is also carefully chosen. Figure 4.6 shows the spatiotemporal variances of the point distributions in four time slots.

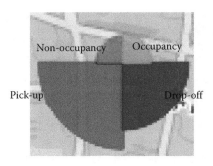

FIGURE 4.5
A pie radar glyph.

From the compact visualization results in Figure 4.6, we can easily observe distinct temporal patterns. The dominant red sectors in Figure 4.6a reflect relatively active pick-ups at 3–4 h, while Figure 4.6b shows several significantly dense drop-off hotspots at 6–7 h. The largest three blue sectors on the west, south, and middle (from left to right) are located in the Hongqiao

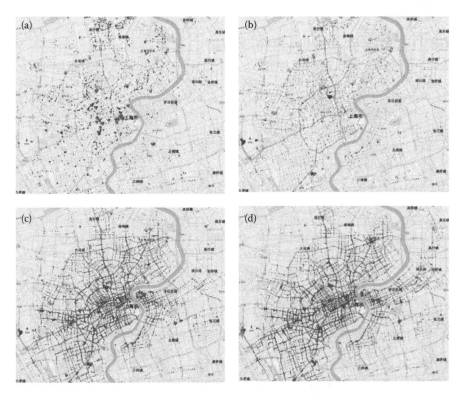

FIGURE 4.6
Temporal distributions of (O, N, P, D) points using the pie radar glyph. (a) 3–4 h. (b) 6–7 h. (c) 8–9 h. (d) 18–19 h.

International Airport (Terminals 2 and 1), Shanghai South Railway Station, and Shanghai Railway Station. Figure 4.6c, d illustrates daily rush hours at 8–9 and 18–19 h, which have a relatively larger number of points. The main roads in the city center are dominantly related to occupancy points. During the rush hour at 8–9 h, there are obvious places of blue sectors of drop-offs and the red sectors of pick-ups scattering around. By contrast, during the rush hour at 18–19 h, there are obvious places of red sectors of pick-ups. Besides, there is a big green sector of non-occupancy points in the airport.

4.3.2 Salient Feature Image

Pie-radar glyph visualization provides easily perceivable overall compact visualization results; however, it still requires high cognitive efforts and interactive operations of the users to explore and understand cluttered symbols. To reduce the visual complexity, we propose a salience-based visualization approach, which displays in each cell (i.e., a spatial partition) only the most salient feature, as opposed to showing all co-located ones.

Taking two time intervals 7–8 and 18–19 h of pick-up and drop-off as exemplary features, the spatial distribution of these two features in the two intervals is demonstrated in Figure 4.7. From the two images in Figure 4.7, we can easily observe dense pick-up (red) and drop-off (blue) areas at the two intervals. The salient features in both maps form roughly complementary spatial distribution patterns. For instance, many dense drop-off areas at 7–8 h become dense pick-up areas at 18–19 h.

Furthermore, in order to display temporal change patterns, we combine the two images into one temporal change map by composing the corresponding pixels of two images. With regard to the salience-based images of pick-ups and drop-offs at 7–8 and 18–19 h, we define three categories of

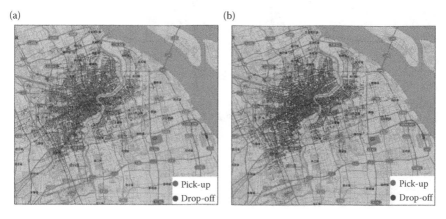

FIGURE 4.7
The salience-based spatial distribution of pick-up and drop-off points. (a) 7–8 h. (b) 18–19 h.

temporal changes. The first category (drop-off ≥ pickup) is the change of the salient feature from drop-off at 7–8 h to pick-up at 18–19 h. Similarly, the second category is (pickup ≥ drop-off). In the last category, the salient feature remains unchanged.

Figure 4.8 shows the temporal change image of the pick-ups and drop-offs at 7–8 and 18–19 h, in which opacity values correspond to the changes in the values of two salient features. It allows a straightforward detection of change patterns during the corresponding time intervals. Regarding the spatial extent, the dominant temporal change (in steel blue) is the change of the salient variable from pick-ups to drop-offs, occupying almost the whole Shanghai area. The second spatially large temporal change is the change from drop-offs to pick-ups, which can be identified by several orange areas. Finally, we can detect quite a few relatively small but shiny yellow areas scattered in Shanghai, which indicates that these areas are of either significant pick-up or drop-off changes. Being stimulated from the spatiotemporal pattern in Figure 4.8 and based on the knowledge of human daily activity patterns, we naturally conjecture that the temporal change patterns are associated with distinct land use types. For instance, areas with intense drop-off activities at 7–8 h and intense pick-ups at 18–19 h probably correspond to the working places. This may be an important supporting indicator for the research work on land use/cover change detection.

To confirm our assumptions, we enclose orange and yellow areas with irregular polygons and manually check these areas on the Shanghai base map. Figure 4.9 shows the distinct areas of interest labeled with their functions or name. The irregular polygons with orange frames and labels are

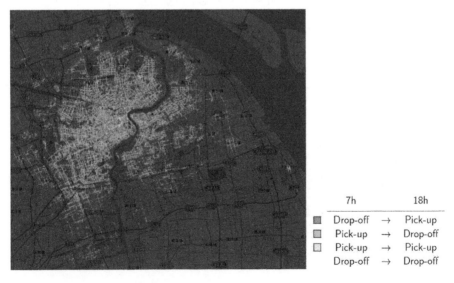

	7h		18h
■	Drop-off	→	Pick-up
■	Pick-up	→	Drop-off
□	Pick-up	→	Pick-up
	Drop-off	→	Drop-off

FIGURE 4.8
The salience change map.

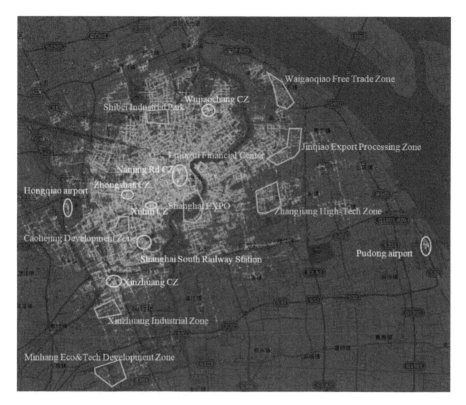

FIGURE 4.9
Regions of interests (CZ stands for commercial zone).

regions especially active with drop-offs at 7–8 h and pick-ups at 18–19 h. We find that most of them are free-trade zones, high-tech zones, industrial and development zones, and financial centers. This observation coincides with our conjecture. Another interesting observation is that the area of Shanghai Expo 2010, located along both banks of the Huangpu River, also belongs to this kind of temporal change pattern. Since Expo 2010 was held from May 1 to October 31, 2010 and covered the total time span of our data set, it is reasonable to expect that the data during this time period can reveal the human mobility patterns related to this international event.

4.4 Trajectory-Based Visual Analysis

In this section, we propose a visual analytics framework for trajectory analysis and apply this framework for the visual analysis of origin–destination lines and non-occupancy trajectories.

4.4.1 A Visual Analytics Framework for Trajectory Analysis

The proposed framework basically consists of three components: (1) visual querying of the movement database, (2) interactive clustering, and (3) visual representations. The framework is implemented in a web-based interactive environment.

1. *Movement Database Visual Querying*: This component allows a visual query of the movement database for interesting trajectory subsets. A computationally efficient way of inspecting massive data is to retrieve only the relevant interesting data partitions from the database. For instance, we can retrieve relevant data sets according to the taxi drivers' income in specific time intervals by interactively brushing an income histogram (Figure 4.10a) and a time line graph (Figure 4.10b).

2. *Interactive Visual Clustering*: This component introduces the interactive visual clustering designed for trajectory exploration. In general, meaningful clustering results require a proper setting of clustering features and parameters, especially for multivariate clustering. For instance, analysts might be interested in inspecting groups of trajectories according to some attribute values, for example, starting from the same locations, or duration less than 1 hour. Furthermore, they may need to examine distinct spatial interaction patterns by setting different distance values. In our study, we design an interactive

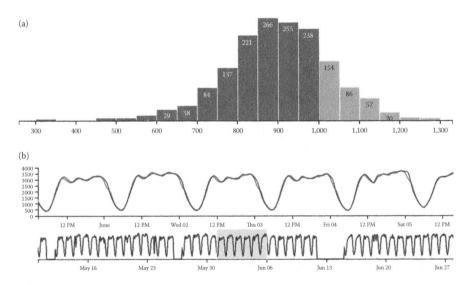

FIGURE 4.10
Visual querying of the movement database. (a) Visual query of taxi drivers' income. (b) Visual query of the time line view.

Clustering Parameters

Feature selection:
☐ distance ☐ duration ☐ start_lon ☐ start_lat ☑ end_lon ☑ end_lat

Distance threshold (meter):

FIGURE 4.11
Interactive clustering interface for feature selection and distance setting.

clustering interface (shown in Figure 4.11) to serve the purpose for feature selection using a checklist and parameter setting with a slide bar.

3. *Visual Representations*: The resulting clusters are visualized in a variety of representations. First, we propose a well-designed parallel coordinates plot to visualize the precomputed clusters resulting from the interactive clustering process. The users can visually detect and interactively select interesting clusters in the parallel coordinates for further analysis. Second, the individual trajectories in the selected clusters are visualized on linked map views. For visualizing origin–destination lines on the map view, we propose a gradient rendering technique, while for non-occupancy trajectories we propose a direct rendering technique and apply the space-time-cube method.

4.4.2 Visual Analysis of Origin–Destination Lines

In this section, we apply the proposed visual analytics framework for origin–destination line analysis.

4.4.2.1 Interactive Hierarchical Agglomerative Clustering

We firstly apply a hierarchical agglomerative clustering method to group the origin–destination lines. An origin–destination line can be modelled as a point in a high-dimension space. The location of the origin and destination, the duration, and distance are used as exemplary attributes. Besides, the algorithm has two parameters, namely, a distance function and a linkage criterion. Figure 4.11 illustrates an example of clustering setting with the selected features of the longitude and latitude of the destination, the clustering distance of 100 m, and an average linkage criterion.

4.4.2.2 A Parallel Coordinates View

We adopt the interactive parallel coordinates technique and design the visual representation of the clustering results as follows. First, besides the multiple attributes (e.g., origin, destination, distance, duration), we add to the parallel coordinates two more features, that is, the individual cluster identity number and the number of elements in each cluster. Second, we design the parallel coordinates in an easily understandable way by ordering the objects using the Z-index and rendering the clusters with semi-transparency. Finally, for an immediate perception of individual clusters, we assign distinctive colors to clusters with a distinct number of cluster elements to reveal the natural clusters.

Taking the trajectories from 7 to 10 h on May 31, 2015 as an example, we obtain about 250 clusters after the interactive clustering with parameter settings in Figure 4.11. These 250 clusters are then represented by the parallel coordinates shown in Figure 4.12a. The first axis represents the number of elements in each cluster. The second axis shows the identity number of each specific cluster. The third and fourth axes represent the distance and duration of the origin–destination lines. The last four axes are for the locations of origin and destination.

As shown in Figure 4.12a, the parallel coordinates reveal the natural data distribution and the clusters in an intuitive manner. Large clusters with more than 15 elements of individual origin–destination lines are colored according to the chosen categorical color scheme. Since we cluster the lines using their destinations, it is natural that the large clusters converge at the last two axes. One more interesting pattern is that larger clusters converge at the middle range of the last two axes, which indicates that large clusters of origin–destination lines happen in the center of the study area. Looking at the first axis (count_elements), we can get an overview about the distribution of the number of elements in each large cluster. For instance, the largest

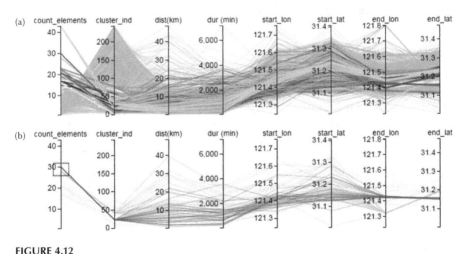

FIGURE 4.12
The parallel coordinates view. (a) Clusters of trajectories based on "destination" from 7 to 10 h on May 31, 2010. (b) Selection of interesting individual clusters.

cluster has approximately 40 elements, and the second 30 elements. For further exploration and analysis, the users can inspect interesting individual clusters by brushing any axis or multiple axes at the same time. As shown in Figure 4.12b, two elements in the first axis are brushed, which correspond to two clusters of about 30 elements.

4.4.2.3 A Map View of Origin-Destination Lines by Gradient Line Rendering

To allow the inspection of the spatial patterns of the selected clusters, we visualize the individual origin–destination lines in a map view, which is linked to the parallel coordinates.

A gradient line rendering technique is proposed to allow an intuitive interpretation of the origin–destination lines. We firstly round the coordinate values of origins/destinations to reduce the line intersections. Then we order the origin–destination lines according to their distances by pushing long lines into the background (using a small z-index) so that short lines would not be hidden. Finally, we divide the lines into segments and assign the series of line segments from origin to destination with gradient colors from dark to light color values.

Taking the selected clusters in Figure 4.12b as an example, Figure 4.13 shows the individual origin–destination lines on the map view after applying the

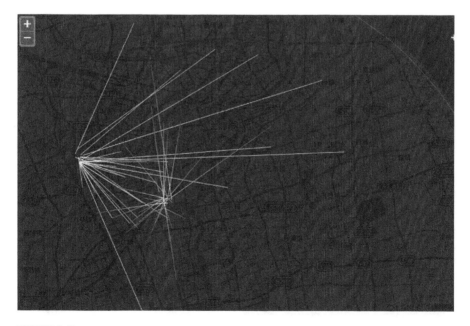

FIGURE 4.13
The map view of the selected origin–destination lines.

gradient rendering technique. The detailed spatial interaction can be clearly identified. The two clusters correspond to two transport hubs with the right-sided cluster of the Shanghai South Railway Station and the left-sided one of the Hongqiao International Airport.

Furthermore, we can interactively select other interesting clusters in the parallel coordinates view and inspect their spatial patterns on the map view. For instance, we brush large clusters in Figure 4.12a and show the individual origin–destination lines in these clusters in Figure 4.14. The largest two clusters of about 42 individual elements (Figure 4.14a) are with their destinations concentrated in the city center. This phenomenon is reasonable since there should be more taxis going to the city center during rush hours. The third and fourth largest clusters of about 30 individual items (Figure 4.14b) correspond to Hongqiao Airport and Shanghai South Railway Station. Owing to the locations of the two transport hubs, especially Hongqiao Airport, there are some long distance origin–destination lines. The fifth to ninth largest clusters (Figure 4.14c, d) are of around 25–30 elements and primarily with destinations near the city center. The gradient line rendering results show not only the spatiotemporal distribution of the lines but also their spatial interaction between different areas. Mostly, there are more local interactions. The radiation shape of the origin–destination lines is different and difficult to foresee, since it relies on the spatial location of the cluster. Users can most

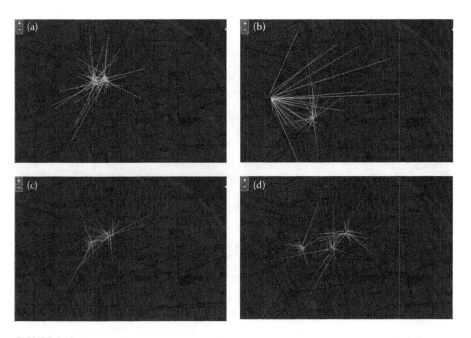

FIGURE 4.14
Significant clusters of destinations at 7–10 h on May 31, 2010. (a) Clusters 1 and 2. (b) Clusters 3 and 4. (c) Clusters 5 and 6. (d) Clusters 7, 8, and 9.

likely explore the spatial patterns of other clusters by brushing their corre-
spondences in the parallel coordinates view.

4.4.3 Visual Analysis of Non-Occupancy Trajectories

In this section, we apply the proposed framework for non-occupancy trajectory analysis of two groups of taxi drivers, that is, high-income and low-
income groups.

Similar to the procedure in Section 4.4.2, we firstly retrieve non-occupancy
trajectories from the movement database by visually querying the low- and
high-income taxis. Then we interactively cluster the trajectories by using the
clustering interface. The clusters are visualized in the parallel coordinates
view. The users can then select significant clusters in the parallel coordinates
and render the individual cluster elements on a 2D map view. Since the characteristics of non-occupancy trajectories are different from origin–destination
lines, we apply the direct line rendering technique for the trajectory visualization. Furthermore, to inspect the space-time dynamic profiles of the trajectories at certain hotspots, we apply the space-time-cube technique.

4.4.3.1 A Map View of Non-Occupancy Trajectories by Direct Line Rendering

To demonstrate, we retrieve the non-occupancy trajectories of the bottom-
and top-performing taxi groups from 6 to 12 h on May 31, 2010. On the basis
of their origins, we cluster these trajectories and show the clustering results
in the parallel coordinates in the upper subfigures in Figure 4.15. Then we
select the significant clusters in the respective parallel coordinates, which are
rendered on the 2-D map with an opacity of 0.2 and with the same colors of
the clusters in the parallel coordinates (the lower subfigures in Figure 4.15).
In the parallel coordinates, we can see that there are larger clusters of non-
occupancy trajectories in the high-income taxi group (Figure 4.15b) than in
the low-income group (Figure 4.15a). In the map views, the overall spatial
distribution of the selected clusters and the driving routes starting from the
cluster centers can be easily detected. The denser line areas are with more
frequent non-occupancy trajectories. We also observe that in each cluster
there is a distance decay effect of the frequency from its center to its border.
Moreover, in terms of the spatial distribution, the cluster centers of both taxi
groups largely overlap. There are also some notable differences. For instance,
in the high-income taxi group, there is a significant cluster related to the
Pudong Airport (on the easternmost), which does not appear in the low-
income group.

By comparing cluster centers with the base map, we can identify the most
important transport hubs in the test area, including Shanghai Railway
Station, Shanghai South Railway Station, Hongqiao Airport, and Pudong
Airport. Figure 4.16 annotates these transport hubs by their names.

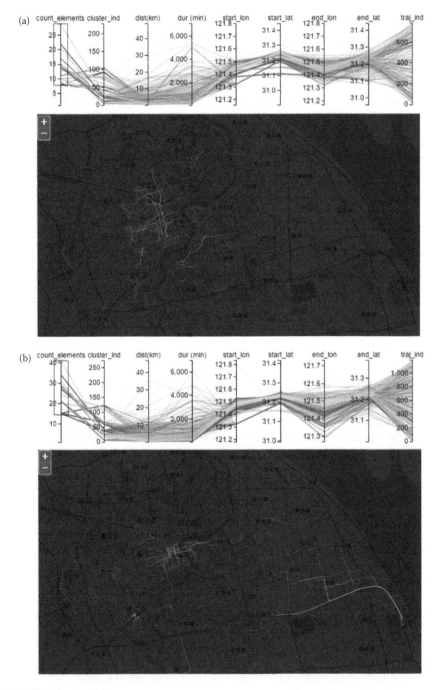

FIGURE 4.15
Non-occupancy trajectory clusters based on "starting location." (a) Bottom-income group. (b) High-income group.

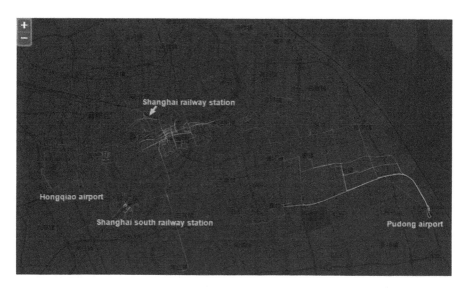

FIGURE 4.16
Identified transport hubs.

4.4.3.2 Space-Time-Cube Visualization of Selected Non-Occupancy Trajectories

To observe the dynamic behavior of individual trajectories at certain hotspots, we visualize them in a space-time cube. We use Pudong International Airport as a test case. For each taxi group, we extract the non-occupancy trajectories starting from or ending at the airport from the movement database. Figure 4.17 shows the dynamics of the non-occupancy trajectories related to the airport on May 31, 2010 in the space-time cubes.

From Figure 4.17, we can get an overview of the spatiotemporal profiles of the trajectories related to the airport. Regarding the number of trajectories, obviously for the top-performing taxi group, there are far more non-occupancy trajectories from the airport (Figure 4.17c) than the ones to the airport (Figure 4.17d). The large number of non-occupancy trajectories from the airport indicates that most of the high-income taxi drivers directly cruise back to the city center after dropping off passengers in the airport rather than waiting there for the next passengers; while the small amount of non-occupancy trajectories to the airport indicates that only a few high-income taxi drivers cruise to the airport without passengers. By contrast, for the bottom income taxi group, we cannot find such a difference.

Looking at the temporal dimension, we can also observe the frequency distribution patterns. For instance, in the bottom performing taxi group, there are more non-occupancy trajectories to Pudong Airport (Figure 4.17b) happening in the afternoon around 15 h, while in the top income taxi group there are more non-occupancy trajectories from the airport (Figure 4.17c) in the early morning (0–5 h).

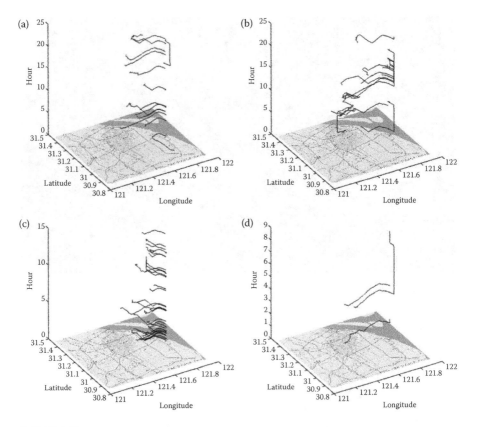

FIGURE 4.17
The space-time cube of non-occupancy trajectories related to Pudong Airport. (a) Bottom-income group from Pudong. (b) Bottom-income group to Pudong. (c) Top-income group from Pudong. (d) Top-income group to Pudong.

Furthermore, we can identify from the non-occupancy trajectories to the airport in both taxi groups (Figure 4.17b, d) that there are many vertical lines. These vertical lines reflect that the taxis are stationary at particular locations for rather long periods. Indeed, we checked these places on the base map and found that they correspond to airport taxi waiting pools where taxi drivers can rest or wait for picking up their next passengers.

4.5 Conclusion

In this chapter, we introduced the state of the art of visual analysis of FCD and our extensive experiments with massive and complex real-world taxi FCD. We analyzed FCD at two abstract levels, namely, point-based and trajectory-based levels. For the visual analysis of multivariate points, that is, occupancy, non-occupancy, pick-up, and drop-off, we proposed a pie-radar glyph method and a salience-based visualization method to reveal their spatiotemporal patterns. The compact visualizations revealed the underlying interesting data distributions, which can be used for further mining, like inferring urban land use types. For the visual analysis of trajectories, we proposed a framework integrating database visual querying, interactive clustering, parallel coordinates, gradient and direct line rendering techniques, and space-time cube, which allows users to freely explore potentially interesting clusters. The framework was applied to visually analyze the spatiotemporal patterns of origin–destination lines and non-occupancy trajectories, which facilitate the understanding of movement interactions.

Acknowledgment

We sincerely thank Professor Chun LIU from Tongji University, Shanghai, for sharing with us the floating car data.

References

Andrienko, G., and N. Andrienko. 2010. A general framework for using aggregation in visual exploration of movement data. *The Cartographic Journal* 47 (1): 22–40. doi: 10.1179/000870409X12525737905042.

Andrienko, G., N. Andrienko, C. Hurter, S. Rinzivillo, and S. Wrobel. 2011. From movement tracks through events to places: Extracting and characterizing significant places from mobility data. *2011 IEEE Conference on Visual Analytics Science and Technology (VAST)*, October 23–28, 2011, Providence, Rhode Island.

Andrienko, G., N. Andrienko et al. 2009. Interactive visual clustering of large collections of trajectories. *2009 IEEE Symposium on Visual Analytics Science and Technology*, October 12–13, 2009, Atlantic City, New Jersey.

Andrienko, N., and G. Andrienko. 2011. Spatial generalization and aggregation of massive movement data. *IEEE Transactions on Visualization and Computer Graphics* 17 (2): 205–219. doi: 10.1109/TVCG.2010.44.

Boyandin, I., E. Bertini, P. Bak, and D. Lalanne. 2011. Flowstrates: An approach for visual exploration of temporal origin-destination data. *Proceedings of the 13th Eurographics/IEEE—VGTC Conference on Visualization*, Bergen, Norway.

Ding, L. 2016. Visual analysis of large floating car data—A bridge-maker between thematic mapping and scientific visualization. PhD thesis, Chair of Cartography, Technischen Universität München, Munich, Germany.

Ding, L., H. Fan, and L. Meng. 2015. Understanding taxi driving behaviors from movement data. In *AGILE 2015: Geographic Information Science as an Enabler of Smarter Cities and Communities*, edited by F. Bacao, M. Y. Santos and M. Painho, 219–234. Cham: Springer International Publishing.

Ding, L., J. Yang, and L. Meng. 2015. Visual analytics for understanding traffic flows of transportation hub from movement data. *Proceedings of the 27th International Cartographic Conference (ICC)*, Rio de Janeiro, Brazil.

Guo, D. 2009. Flow mapping and multivariate visualization of large spatial interaction data. *IEEE Transactions on Visualization and Computer Graphics* 15 (6): 1041–1048. doi: 10.1109/TVCG.2009.143.

Guo, D. and X. Zhu. 2014. Origin-destination flow data smoothing and mapping. *IEEE Transactions on Visualization and Computer Graphics* 20 (12): 2043–2052. doi: 10.1109/TVCG.2014.2346271.

Guo, H., Z. Wang, B. Yu, H. Zhao, and X. Yuan. 2011. TripVista: Triple perspective visual trajectory analytics and its application on microscopic traffic data at a road intersection. *2011 IEEE Pacific Visualization Symposium*, Hong Kong.

Holten, D. 2006. Hierarchical edge bundles: Visualization of adjacency relations in hierarchical data. *IEEE Transactions on Visualization and Computer Graphics* 12 (5): 741–748. doi: 10.1109/TVCG.2006.147.

Holten D., and J. J. Van Wijk. 2009. Force-directed edge bundling for graph visualization. *Computer Graphics Forum* 28 (3): 983–990. doi: 10.1111/j.1467-8659.2009.01450.x.

Keim, D., J. Kohlhammer, G. Ellis, and F. Mansmann, eds. 2010. *Mastering the Information Age—Solving Problems with Visual Analytics. Eurographics Association*, Geneva, Switzerland.

Keler, A., L. Ding, and M. J. Krisp. 2016. Visualization of traffic congestion based on Floating Taxi Data. *Kartographische Nachrichten* 66 (1): 7–13.

Kraak, M.-J. 2003. The space-time cube revisited from a geovisualization perspective. *Proceedings of the 21st International Cartographic Conference*, Durban, South Africa.

Liu, L., C. Andris, and C. Ratti. 2010. Uncovering cabdrivers' behavior patterns from their digital traces. *Computers, Environment and Urban Systems* 34 (6): 541–548. doi: 10.1016/j.compenvurbsys.2010.07.004.

Liu, X., L. Gong, Y. Gong, and Y. Liu. 2015. Revealing travel patterns and city structure with taxi trip data. *Journal of Transport Geography* 43: 78–90. doi: 10.1016/j.jtrangeo.2015.01.016.

Liu, Y., X. Liu et al. 2015. Social sensing: A new approach to understanding our socioeconomic environments. *Annals of the Association of American Geographers* 105 (3): 512–530. doi: 10.1080/00045608.2015.1018773.

Mazimpaka, J. D. and S. Timpf. 2015. Exploring the potential of combining taxi GPS and flickr Data for discovering functional regions. In *AGILE 2015: Geographic Information Science as an Enabler of Smarter Cities and Communities*, edited by F. Bacao, M. Y. Santos and M. Painho, 3–18. Cham: Springer International Publishing.

Michal, P., S.-D. Natasa, and G. Matthias. 2009. CRAWDAD dataset epfl/mobility (v. 2009-02-24).

Santi, P., G. Resta et al. 2014. Quantifying the benefits of vehicle pooling with shareability networks. *Proceedings of the National Academy of Sciences* 111 (37): 13290–13294.

Scheepens, R., N. Willems et al. 2011. Composite density maps for multivariate trajectories. *IEEE Transactions on Visualization and Computer Graphics* 17 (12): 2518–2527. doi: 10.1109/TVCG.2011.181.

Thomas, J. J. and K. A. Cook, eds. 2005. *Illuminating the Path: The Research and Development Agenda for Visual Analytics.* National Visualization and Analytics Center, Richland, WA.

Tominski, C., H. Schumann, G. L. Andrienko, and N. V. Andrienko. 2012. Stacking-based visualization of trajectory attribute data. *IEEE Transactions on Visualization and Computer Graphics* 18 (12): 2565–2574. doi: 10.1109/TVCG.2012.265.

van den Elzen, S. and J. J. van Wijk. 2014. Multivariate network exploration and presentation: From detail to overview via selections and aggregations. *IEEE Transactions on Visualization and Computer Graphics* 20 (12): 2310–2319. doi: 10.1109/TVCG.2014.2346441.

Wong, P. C. and R. D. Bergeron. 1997. 30 years of multidimensional multivariate visualization. *Scientific Visualization, Overviews, Methodologies, and Techniques,* pp. 3–33, IEEE Computer Society, Washington, DC.

Wood, J., J. Dykes, and A. Slingsby. 2010. Visualisation of origins, destinations and flows with OD maps. *The Cartographic Journal* 47 (2): 117–129. doi: 10.1179/00087 0410X12658023467367.

Yuan, J., Y. Zheng, and X. Xie. 2012. Discovering regions of different functions in a city using human mobility and POIs. *Proceedings of the 18th ACM SIGKDD International Conference on Knowledge Discovery and Data Mining,* pp. 186–194, Beijing, China. doi: 10.1145/2339530.2339561.

Zheng, Y., L. Zhang, X. Xie, and W.-Y. Ma. 2009. Mining interesting locations and travel sequences from GPS trajectories. *Proceedings of the 18th International Conference on World Wide Web,* pp. 791–800, Madrid, Spain. doi: 10.1145/1526709.1526816.

Zhou, H., P. Xu, X. Yuan, and H. Qu. 2013. Edge bundling in information visualization. *Tsinghua Science and Technology* 18 (2): 145–156. doi: 10.1109/TST.2013.6509098.

5

Recognizing Patterns in Geospatial Data Using Persistent Homology: A Study of Geologic Fractures

Kathryn A. Bryant and Bobak Karimi

CONTENTS

5.1 Introduction

The goal of this chapter is to describe a mathematical tool that can be used to recognize patterns of similarity in data. We will be interested in how this tool applies to geospatial data, but it is important to note that it can be applied to any data that can be described numerically, and is most directly applicable to vector data types. Raster data types can also be processed by such a tool, but only after processing pixel areas to point or other vector data types. To explore the efficacy of such a tool, and since the tool is in its early stages, we will use the tool to explore synthetic point data sets that we relate to geologic structures at a tectonic boundary.

The tool that this chapter aims to describe comes from the mathematical field of topology. In imprecise terms, topology is the study of mathematical spaces and their characteristics that are independent of size/distance. To illustrate what this means, we first contrast the notion of a topological property with the more familiar notion of a geometric property, such as area, volume, arc length, or curvature. Each of these is a property we can compute for a space (like a polygon or a solid) that depends on the size of the object in question. We know, for example, that two spheres with different length radii will have different volumes. As such, volume (and area, arc length, curvature, etc.) is a geometric property rather than a topological one. With these geometric properties in mind for contrast, examples of topological properties for a space are the following:

- *Number of components*: Does the space have one or three pieces?
- *Compactness**: Does the space extend infinitely and/or have a boundary?
- *Number/shape of holes*: Does the space resemble a basketball, baseball, an inner tube, or a donut?

In what follows, we will focus on this last topological property of number of holes and types of holes in a space, which are detected by a tool called *homology*.

Homology is able to rigorously distinguish an inner tube from a donut by capturing the fact that while an inner tube and a donut both have a common central hole, an inner tube has an additional hole (sweeping around the other hole) into which a donut could fit (again, ignoring size!). Using an advanced mathematical tool like homology to rigorously distinguish an inner tube and a donut may seem silly, but only because we can actually *see* donuts and inner tubes. The utility of homology becomes much greater and more apparent when we start trying to understand and distinguish mathematical objects in higher dimensions where our ability to visualize the spaces in question is lost. Since our goal is to apply persistent homology (a variation of traditional homology) to data, it is crucial to note that the concept of "higher dimensions" is not merely theoretical; if we look at a data set each of which have, say, six attributes, then our data set is a 6-dimensional object. Hence, having a tool that detects topological features of a space in any dimension and that does not at all rely on human visualization abilities is extremely useful.

The kinds of spaces on which homology is computed tend to be *manifolds*, which are spaces that "look like" Euclidean space of some dimension—at

---

* In mathematics, the definition of a compact space is a technical one. For a rigorous definition, see Munkres (1975).

least in small neighborhoods in the space around each point. For example, we would consider a balloon (ignoring the tie) a 2-dimensional manifold because if we zoom in very, very close to any single point on the balloon, we can be tricked into thinking that it is a plane, that is, a 2-dimensional Euclidean space. Alternatively, if we were to describe a balloon (a sphere) with equations, we would have two degrees of freedom in the equations and therefore define a 2-dimensional manifold. The fact that a balloon needs three dimensions to exist in as a whole has no bearing on how we classify its dimension as a manifold. Thus, since very tiny neighborhoods around any point in a balloon look like a 2-dimensional Euclidean space, we regard a balloon as a 2-dimensional manifold (also called a *surface*). For a more rigorous definition of a manifold, see Guillemin and Pollack (2010).

In looking forward to computing the homology of a data set, we encounter a problem: while it is perfectly reasonable to compute the homology of a discrete space (a space with finitely many points) like a data set or point cloud, the fact that single points and collections of single points are regarded as 0-dimensional manifolds means that there are not any interesting "holes" to measure because the dimension is too small. At best, homology would merely tell us how many points are in the space, which is not a bit of information worth the trouble of employing this mathematical tool. To get homology to capture more interesting features of our data, we will turn our data sets—our 0-dimensional manifolds—into objects of higher dimension. We will do this by first defining a *metric* on our data that numerically captures similarity between any pair of points in our data set; next, we will continuously build a *simplicial complex* from our data set using the aforementioned metric, thereby transforming our set of single points into a space with higher dimension; then lastly, we will compute the homology of the corresponding simplicial complex. Since homology detects information about holes in spaces, this computation will tell us about holes in our data, where "holes" in this case correspond to interesting types of clustering. For example, if points in our data set are clustered around a circle, then computing the homology of an appropriately scaled simplicial complex built from our data will reveal this clustering.

It should be noted that clustering of data points around circles, tori, spheres, and other interesting mathematical objects is a phenomenon that cannot be detected using statistics, the go-to analytic discipline for applications. In mathematics, *spheres* are "hollow" objects, defined to be the set of points exactly some distance r from a center point. It is actually redundant to mention circles and spheres in the same list because a circle is a 1-dimensional sphere, denoted by S^1; in general, a k-dimensional sphere is denoted by S^k. On the other hand, *balls* are "solid" objects, defined to be the set of points less than or equal to some distance from a center point. A k-dimensional ball is denoted by B^k of D^k. Here, we define the dimension of a sphere or a ball to agree with these objects' dimensions as manifolds. As a result, the boundary of a k-dimensional ball is always a $(k-1)$-dimensional sphere. For example, the 2-ball is a solid disk whose boundary is the circle, that is, the 1-sphere. Statistics can only detect clustering inside a ball, and therefore fails to distinguish clustering around a circle versus clustering inside a 2-dimensional disk. The ability of homology to detect these more unusual kinds of clustering is one of the main arguments for using topology as an additional tool to study data.

5.2 Metric Spaces and Simplices

Geospatial data often come equipped with two or three coordinates that describe the physical locations of the data points. With these physical coordinates, we can get visuals of the locations of our data points and judge them, with our Euclidean sense of distance, to be distributed or clustered in certain ways. However, rarely are spatial attributes the only attributes we care about in geospatial data. As an example, in linear data sets, the data set might contain attributes describing the line length and/or azimuth in addition to coordinates describing the data points' 2- or 3-dimensional locations. When we attempt to detect patterns in linear sets, these extra attributes can play a role in determining which points are similar.

The way to accomplish this sort of similarity-detection mathematically is to turn a data set into a *metric space*, which will allow us to account for the spatial distance between data points *and* other attributes of importance. A metric space is a set of objects together with a well-defined notion of distance between each pair of objects in the set. Abstractly, if A is a set, then a "notion of distance" on A is a function d, in the mathematical sense, that takes two members of A as inputs and yields a non-negative real number as an output representing the distance between the two (inputted) objects in the set.

The most familiar metric spaces are Euclidean metric spaces, in which the set of objects A is the set of all points in some Euclidean space, such as the real line $\mathbb{R}$, the plane $\mathbb{R}^2$, or usual 3-space $\mathbb{R}^3$, and the distance between any pair of objects in these spaces is given by the distance formula. For the plane $\mathbb{R}^2$, for example, the distance between two points $a = (x_1, y_1)$ and $b = (x_2, y_2)$ is given by the function $d(a,b) = \sqrt{(x_2 - x_1)^2 + (y_2 - y_1)^2}$. This particular distance function and its higher dimensional analogs are collectively called Euclidean Metrics.

It is possible, however, to consider the same set A of data points in a Euclidean space but under a different notion of distance between points. Using $\mathbb{R}^2$ again as an example, we could instead use the "taxicab metric" to determine distance between two points. Under the taxicab metric, the distance between two points $a = (x_1, y_1)$ and $b = (x_2, y_2)$ is given by $d_T(a,b) = |x_1 - x_2| + |y_1 - y_2|$. This metric is so named because, rather than compute the distance between two points as the length of a straight line segment between them, we imagine that only strict horizontal and vertical line segments exist between points, as for a taxi driving between two points and having to follow city streets; this determines the distance between two points to be the smallest of all possible sums of the lengths of horizontal line segments and vertical line segments connecting the two points, that is, the shortest route for a taxi between two points.

The purpose of looking at the Euclidean metric and the taxicab metric on the same set is to illustrate that different notions of distance can exist on the same set (and naturally, we might add) and to motivate the more general notion of a metric space. Given a space of objects A, we are permitted to define a distance function d on A in any way we choose, provided that the distance function satisfies the following: For objects a, b, and c in A,

1. (Nonnegativity) $d(a,b) \geq 0$
2. (Definiteness) $d(a,b) = 0$ if and only if $a = b$
3. (Symmetry) $d(a,b) = d(b,a)$
4. (Triangle Inequality) $d(a,c) \leq d(a,b) + d(b,c)$.

These four requirements are precisely the ones we need to create distance functions that obey our usual intuition of how a distance function should behave. We require non-negativity so that we avoid negative distances; we require definiteness so that the only circumstance under which we get a distance of zero between two objects is when those two objects are actually the same object; we require symmetry so that the distance between two objects does not depend on which one we "start" at; and lastly, we require the triangle inequality to ensure that adding an extra point along a route never reduces the total distance traveled. For more information on metric spaces, see Kumaresan (2005).

Given a data set A, we can turn the data set into a metric space by defining a distance function d on A that captures similarity and dissimilarity of data points (observations) based on their attributes of importance. We will define a metric so that two data points are "close" if they are similar across all attributes. If the observations in a data set contain spatial coordinates as attributes (as will often be the case with geospatial data), we can plot these data points using their spatial coordinates and visually communicate any additional similarity by connecting points that are metrically close with *simplices*.

An *n-simplex* is the convex hull of $n+1$ distinct points, meaning that it is the smallest convex set that contains the chosen $n+1$ points. Examples of *n*-simplices for the first few values of n are given in Figure 5.1. A 0-simplex is merely a point or *vertex*; a 1-simplex is a line segment or *edge*; a 2-simplex is a solid triangle; and a 3-simplex is a solid tetrahedron. Although we cease to give them special names like vertex, edge, triangle, or tetrahedron, *n*-simplices with $n \geq 4$ are covered by the given definition and are simply the higher dimensional analogs of these more familiar objects.

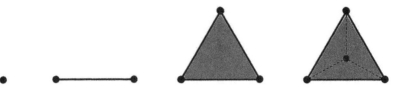

FIGURE 5.1
From left to right: a 0-simplex (vertex), a 1-simplex (edge), a 2-simplex (solid triangle), and a 3-simplex (solid tetrahedron).

We will use simplices to capture similarity between points and store this information in a *simplicial complex*, so named because it can be decomposed into a finite number of simplices.* A simplicial complex is a generalization of a more basic mathematical object called a *graph*. In the mathematical field of graph theory, a graph is an object consisting solely of vertices (points/nodes) and edges (line segments between vertices). A simplicial complex is regarded as a generalization of a graph in that, rather than consist exclusively of vertices and edges as in a graph, a simplicial complex may contain vertices (0-simplices), edges (1-simplices), solid disks (2-simplices), solid balls (3-simplices), and higher-dimensional simplices.

For a data set A with an associated metric d, there are two classical types of simplicial complexes we can create to visually communicate similarity information about the points in A. For a fixed value $r \geq 0$, let $B_r^n(p)$ denote the solid n-dimensional ball of radius r centered at point p. Then, given a finite collection of points in n-dimensional Euclidean space:

1. The *Čech complex* C_r is the simplicial complex in which a k-simplex, defined by unordered $(k+1)$-tuples of points $\{p_0, p_1, \ldots, p_k\}$, is included in the complex if the intersection of all the $B_r^n(p_i)$'s (for $0 \leq i \leq k$) contains a point .

2. The *Rips complex* R_r is the simplicial complex in which a k-simplex, defined by unordered $(k+1)$-tuples of points $\{p_0, p_1, \ldots, p_k\}$, is included in the complex if, for $0 \leq i, j, \leq k$, the intersection of each pair of balls $(B_r^n(p_i), B_r^n(p_j))$ contains a point.

Both Čech and Rips complexes are visuals of data point similarity up to the threshold or *filtration value r*. If the type of complex we are creating at a certain r-value has not been specified, we will call the complex Δ_r. An appropriate or helpful filtration level will depend entirely on the scale(s) at which the observation attributes are measured. Note that if $r = 0$, then Δ_0 will merely be our set of data placed in 2- or 3-space with nothing between them, that is, it will just be a point cloud. On the other hand, if r is enormous (again, relative

---

* Technically, the simplices in a simplicial complex K must also satisfy the following: (i) if σ is a simplex in K and τ is a *face* of σ, then τ is also a simplex in K, and (ii) the intersection of σ and τ is either empty or a face of both (Edelsbrunner and Harer, 2010).

to the scale at which the attributes of our observations were measured), then every point will be connected to every other point and Δ_r reveals no information about similarity.

5.3 Classical and Persistent Homology

5.3.1 Preliminaries: Groups, Boundary Operators, and Chain Complexes

The process of computing homology for a space is a way of translating geometric information about a space into algebraic information. Broadly speaking, algebraic objects are sets with operations defined on them; familiar examples of algebraic objects include the set of real numbers with the operations of addition and multiplication, or the set of n-dimensional vectors with real number entries and the operations of vector addition, dot product, cross product, and scalar multiplication (i.e., a *real vector space*). The geometric object we will start with is a simplicial complex (built as above from a data set with a metric defined on it) and we will encode its relevant geometric information in an algebraic object called a *chain complex*; once we have a chain complex, we will analyze it and form *homology groups,* which are the algebraic objects that will describe the shape of our simplicial complex and thus clustering phenomena in our data.

Homology theories can be defined on mathematical objects other than simplicial complexes; therefore, the version given here is a particular kind of homology called *simplicial homology*. The definition of simplicial homology requires the following definitions:

1. A *chain* of n-simplices is any collection of n-simplices within a simplicial complex. See Figure 5.2.

2. A chain of n-simplices that begins and ends in the "same place" is called a *cycle*. For example, a chain of 1-simplices that forms a polygonal loop is a 1-cycle; a chain of 2-simplices that fit together to create a polygonal "balloon" or 2-sphere is a 2-cycle. See Figure 5.3.

3. In general, for $n \geq 1$, the *boundary* of an n-simplex is a closed cycle of $(n-1)$-simplices. For example, a 2-simplex (a solid triangle) has a 1-cycle (a polygonal circle) as its boundary. See Figure 5.4.

4. A closed cycle of n-simplices that does *not* enclose an $(n+1)$-simplex yields a hole of dimension n. An example to keep in mind is that of a closed cycle of 2-simplices (a polygonal 2-sphere) that does *not* bound a 3-simplex (a solid tetrahedron). See Figure 5.5.

As mentioned, a necessary step in defining simplicial homology is to encode geometric information about a simplicial complex in a *chain complex*.

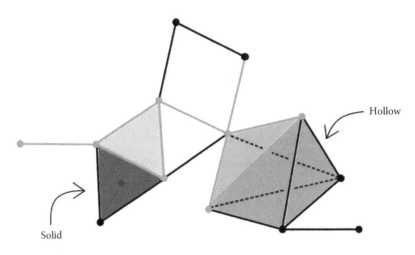

FIGURE 5.2
A chain, highlighted in green.

A chain complex is an ordered sequence of algebraic objects called *abelian groups* in which these groups are connected by special functions called *boundary homomorphisms*, *boundary operators*, or *differentials*. A *group* is a set of objects on which an operation for combining elements is defined so that the combined object is also an object in the set.* For example, the set of integers

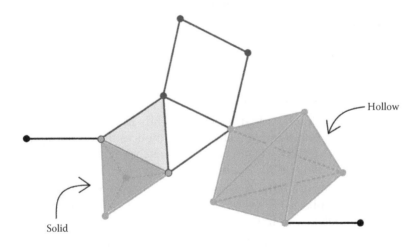

FIGURE 5.3
1-cycle (highlighted in pink) and 2-cycles (highlighted in green). Theorange edge belongs to both the 1-cycle bounding the 2-simplex and the 2-cyclebounding the 3-simplex.

---

* There are actually four group axioms: *closure, associativity, identity, and invertibility*, and we have only addressed closure. To learn about the others, see Gallian (2010).

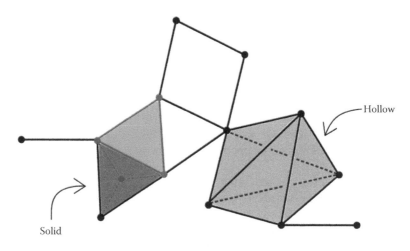

FIGURE 5.4
A 2-simplex (in green) and its boundary (a 1-cycle, in darker green).

under the operation of addition is a group. An *abelian group* is a group that is commutative with respect to the chosen operation; for integers a and b, $a+b=b+a$, and therefore the integers under addition are an abelian group. An abelian group is actually an example of a more general algebraic object called a *module over the integers*, which means that in addition to the group operation, scalar multiplication of a group element by an integer is defined.

The group we choose to use in a chain complex will give us an algebraic structure that we can use to make sense of our simplices, and it will be referred to as the *coefficient group*. The two most commonly used coefficient

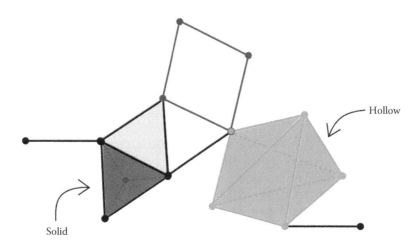

FIGURE 5.5
A 1-dimensional hole (highlighted in pink) and a 2-dimensional hole (highlighted in green).

groups for simplicial homology are the integers under addition, denoted by $\mathbb{Z}$, and the integers modulo 2 under addition, denoted by $\mathbb{Z}/2\mathbb{Z}$ or $\mathbb{Z}_2$. For a complete explanation of modular arithmetic, see Stewart (1995). The integers as an abelian group/module is the set of integers $\{\ldots,-2,-1,0,1,2,3,\ldots\}$ with the usual notions of addition and scalar multiplication; the integers modulo 2 as an abelian group/module over the integers is the set $\{0,1\}$ under the addition rules of $0+0=0$, $0+1=1+0=1$, and $1+1=0$ and the scalar multiple rules of $0\cdot0=0$, $0\cdot1=0$, and $1\cdot1=1$. For these two groups, adding simplices and chains of simplices proceeds as follows:

1. For $\mathbb{Z}$, the simplices in a chain can have any integer coefficient. For example, if σ and τ are two simplices, then $-12\sigma+3\tau$ is an example of a chain. If σ and τ were both 1-simplices, we could think of the chain $-12\sigma+3\tau$ as the chain that includes both σ and τ and for which we traverse σ 12 times "backwards" and τ three times "forwards," under some predetermined notion of forwards and backwards. This concept of direction along a 1-simplex is a special case of the much more general concept of *orientation* that can be defined on a simplex (or manifold) of any dimension. Although we will not elaborate on this, the reader should know that if we were looking at the chain $-12\sigma+3\tau$ where instead σ and τ were 5-simplices, then there would be an analogous notion of "traversing" σ and τ along with or against some orientation just as in the 1-simplex case. For now, this is a formality that can be ignored.

2. For $\mathbb{Z}_2$, the simplices in a chain can only have coefficients of 0 or 1; if the coefficient of a simplex is 0, that term will not be written. Given the rules of addition in $\mathbb{Z}_2$, it also follows that $\sigma+\sigma=(1+1)\sigma=0\cdot\sigma=0$. Thus, every chain over $\mathbb{Z}_2$ is merely a sum of simplices, each of which has coefficient 1. The module $\mathbb{Z}_2$ is used as a coefficient group in simplicial homology when we want to ignore orientation. Since every chain over $\mathbb{Z}_2$ is a sum of simplices with coefficient 1, we ignore the number of times or direction of traversal of each simplex and simply acknowledge its inclusion in the chain. Homology with $\mathbb{Z}_2$ coefficients is a simplified version of homology with integer coefficients because it allows us to view the presence of simplices in a chain as binary: in or out.

Along with the choice of a coefficient group, a chain complex requires a boundary operator. The boundary operator, denoted by ∂, takes n-chains as inputs and outputs $(n-1)$-chains corresponding to the boundary of the inputted n-chain. (This is an aptly named function!). For example, if a and b are 0-simplices (vertices) and σ is the 1-simplex (edge) whose endpoints are a and b, then with $\mathbb{Z}$ coefficients the boundary operator would give us $\partial(\sigma)=a-b$; the first endpoint a is added and the second endpoint b is subtracted to indicate the ordering on the endpoints that indicates our preferred orientation of σ. With $\mathbb{Z}_2$ coefficients, the boundary operator would give us $\partial(\sigma) = a + b$, since both a and b are merely present as boundary elements of σ.

Another name for a boundary operator that we mentioned above is boundary homomorphism. The word *homomorphism* indicates that the map ∂ has the following property: for any objects σ and τ on which ∂ is defined, $\partial(\sigma + \tau) = \partial(\sigma) + \partial(\tau)$. With our homomorphism being one that takes simplices as inputs and outputs the corresponding boundary simplices, the homomorphism property simply says that taking the boundary of a compound chain of simplices (such as $\sigma + \tau$) will produce the same answer as taking the boundary of each simplex in a chain individually and adding the results.

With a chosen coefficient group and the boundary operator in hand, building a chain complex consists of writing down *chain groups* C_n, one for each dimension n, that correspond to the simplices in each dimension. The chain group C_n is formed by taking one copy of the group ($\mathbb{Z}_2$, in our example) for each n-simplex in the simplicial complex and adding (in the sense of groups) all these copies together; k copies of $\mathbb{Z}_2$ added together will be denoted by $(\mathbb{Z}_2)^k$. We then connect the chain groups in neighboring dimensions by the boundary operator, meaning that for the boundary operator ∂ going between C_n and C_{n-1}, the domain of ∂ is C_n and the image of ∂ lies in C_{n-1}. In doing this, we create an algebraic object—the chain complex—that encodes the basic geometric relationships between the simplices in our simplicial complex. For ease of reference, it is common to denote the boundary operator whose inputs are n-simplices and outputs are $(n-1)$-simplices by ∂_n.

5.3.2 Definition: Simplicial Homology

Once the chain complex has been written down for a simplicial complex, we can define simplicial homology. Defining simplicial homology from an associated chain complex relies on the following observations:

1. The domain of each ∂_{n+1} is *all* $(n+1)$-simplices, but the image of ∂_{n+1} is only those n-chains that are boundaries of $(n+1)$-simplices. For example, the domain of ∂_3 consists of all 3-chains, but the output or image of ∂_3 consists of 2 chains that appear in the simplicial complex as the boundary of some 3-simplex. We will denote the set of n-chains that are boundaries of $(n+1)$-simplices by B_n. As a technical note, B_n is actually a *submodule* of C_n, and this is needed later to define the nth homology group.

2. All cycles have "empty" boundaries; therefore, ∂_n sends any n-cycle to the empty $(n-1)$-cycle, that is, to 0. We will denote the set of n-cycles by Z_n. Again, as a technical note, Z_n is a submodule of C_n.

3. Recall that while the boundary of an $(n+1)$-simplex is always an n-cycle, it is *not* true that every n-cycle is the boundary of an $(n+1)$-simplex; for example, a 1-cycle that is not the boundary of

a 2-simplex corresponds to a "hole" of dimension one. Therefore, B_n is actual a submodule of Z_n^* as well, and to capture the holes of dimension n in our simplicial complex, we look at the difference between the set of n-cycles Z_n and the set of n-cycles that are boundaries of $(n+1)$-simplices B_n.

With the above observations, we define simplicial homology:

Definition:

The nth *homology group* $H_n(X)$ of a simplicial complex X is the quotient group $H_n(X) = Z_n(X)/B_n(X)$. $H_n(X)$ is the algebraic object that represents the n-dimensional holes and their topological behavior in X. The X is often suppressed in homological notation when the simplicial complex in question is clear from the context.

For a precise definition of a quotient group, see Gallian (2010). Also, it was noted earlier that Z_n and B_n are subgroups of C_n, which are necessary for H_n to be a group itself.

With the definition of the nth homology group H_n finally in our hands, the remaining task is to execute the computation H_n for each $n \geq 0$. The most streamlined and easily generalizable way to compute homology groups is to write down the boundary operators as linear maps, that is, as matrices, and use the power of linear algebra and computers to actually do the computations. The boundary operator ∂_{n+1}, whose domain is C_{n+1} and whose image lies in C_n, can be written down as a matrix by first choosing the bases for C_{n+1} and C_n to be the $(n+1)$-simplices and the n-simplices in the simplicial complex, respectively. Then, ∂_{n+1} is the matrix whose rows represent the n-simplices and whose columns represent the $(n+1)$-simplices. If there are k $(n+1)$-simplices and m n-simplices (said differently, C_{n+1} has *rank k* and C_n has rank m), then the matrix for ∂_{n+1} has dimension $m \times k$.

Once all of the nonzero boundary operators are written down as linear maps in matrix form, the matrices are then put into *Smith normal form*. For more information on Smith Normal form, see Jäger (2003). With $\mathbb{Z}_2$ coefficients, the Smith normal form of a matrix will have some number of 1's down the main diagonal of the matrix and 0's everywhere else; with $\mathbb{Z}$ coefficients, the Smith normal form will have some number of nonzero integer entries down the main diagonal and 0's everywhere else. For the matrix of the boundary operator ∂_{n+1}, the number of all-zero columns in the Smith normal form of the corresponding matrix gives the rank of Z_{n+1}; the number of nonzero rows in the Smith normal form of the matrix gives the rank of B_n.

---

* This fact, that B_n is a submodule of Z_n, harkens from a fundamental property of the boundary operators, namely, that $\partial \circ \partial = 0$.

It is from this information that we can write down our homology groups. If we denote the rank of Z_n by z_n and the rank of B_n by b_n, then the rank of H_n, called the nth *Betti number* and denoted by β_n, is given by $\beta_n = |z_n - b_n|$.

For $\mathbb{Z}_2$ coefficients, knowing the nth Betti number β_n tells us exactly what the nth homology group is: $H_n = (\mathbb{Z}_2)^{\beta_n}$. The situation is slightly more complicated with $\mathbb{Z}$ coefficients because any integer along the main diagonal of the matrix in Smith normal form will produce something called a *torsion* submodule/subgroup in the corresponding homology group, but Betti numbers are still a decent rough measure of interesting hole-like phenomena in the simplicial complex.

Despite what may seem like a very technical discussion, the above definition and computational algorithm given for simplicial homology omits a significant amount of underlying mathematical theory. A few of the main topics foundational to homology that were not even mentioned above are *homotopy invariance, induced maps,* and *homology classes.* Other aspects that were mentioned, but not in full detail, are the precise definitions of: modules, groups, quotient modules/groups, submodules/subgroups, exact and short exact sequences, bases for modules, ranks of linear maps/dimensions of modules, and matrix reduction and Smith normal form. A recommended, and in-depth, text on homology that addresses all of the aforementioned topics, and also goes on to talk about persistent homology and other applications of topology, is *Computational Topology* by Edelsbrunner and Harer.

5.3.3 Example: Computing Simplicial Homology

We will demonstrate the algorithm for defining and computing homology using the simplicial complex Y in Figure 5.6 with $\mathbb{Z}_2$ coefficients. The reader should keep in mind, however, that everything described can also be done on any simplicial complex and is often done with the coefficient group $\mathbb{Z}$.

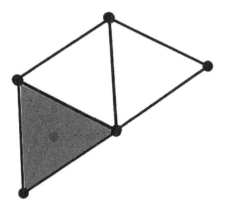

FIGURE 5.6
Simplicial complex Y; note that the tetrahedron formed here is *hollow,* unlike in the earlier example simplex.

In the example simplicial complex Y, there are: zero n-simplices for $n \geq 3$ ($C_{n \geq 3} = 0$); four 2-simplices ($C_2 = (\mathbb{Z}_2)^4$); eight 1-simplices ($C_1 = (\mathbb{Z}_2)^8$); and five 0-simplices ($C_0 = (\mathbb{Z}_2)^5$). As such, the chain complex for Y is

$$\cdots \xrightarrow{\partial_5} 0 \xrightarrow{\partial_4} 0 \xrightarrow{\partial_3} (\mathbb{Z}_2)^5 \xrightarrow{\partial_2} (\mathbb{Z}_2)^{10} \xrightarrow{\partial_1} (\mathbb{Z}_2)^6$$

With the bases for our boundary operators as described above and ordered according to Figure 5.6, we get the following matrices for ∂_2 and ∂_1. The second matrix listed for each map is the Smith Normal Form of the matrix, from which we will compute z_n, b_n, and β_n.

$$\partial_2 = \begin{bmatrix} 1 & 1 & 0 & 0 \\ 0 & 1 & 1 & 0 \\ 0 & 1 & 0 & 1 \\ 0 & 0 & 1 & 1 \\ 1 & 0 & 0 & 1 \\ 1 & 0 & 1 & 0 \\ 0 & 0 & 0 & 0 \\ 0 & 0 & 0 & 0 \\ 0 & 0 & 0 & 0 \\ 0 & 0 & 0 & 0 \end{bmatrix} \sim \begin{bmatrix} 1 & 0 & 0 & 0 \\ 0 & 1 & 0 & 0 \\ 0 & 0 & 1 & 0 \\ 0 & 0 & 0 & 0 \\ 0 & 0 & 0 & 0 \\ 0 & 0 & 0 & 0 \\ 0 & 0 & 0 & 0 \\ 0 & 0 & 0 & 0 \\ 0 & 0 & 0 & 0 \\ 0 & 0 & 0 & 0 \end{bmatrix}$$

$$z_2 = 1, \quad b_1 = 7$$

$$\partial_1 = \begin{bmatrix} 1 & 0 & 1 & 0 & 1 & 0 & 0 & 0 & 0 & 0 \\ 1 & 1 & 0 & 0 & 0 & 1 & 1 & 0 & 0 & 1 \\ 0 & 1 & 1 & 1 & 0 & 0 & 0 & 0 & 0 & 0 \\ 0 & 0 & 0 & 1 & 1 & 1 & 0 & 1 & 0 & 0 \\ 0 & 0 & 0 & 0 & 0 & 0 & 1 & 1 & 1 & 0 \\ 0 & 0 & 0 & 0 & 0 & 0 & 0 & 0 & 1 & 1 \end{bmatrix} \sim \begin{bmatrix} 1 & 0 & 0 & 0 & 0 & 0 & 0 & 0 & 0 & 0 \\ 0 & 1 & 0 & 0 & 0 & 0 & 0 & 0 & 0 & 0 \\ 0 & 0 & 1 & 0 & 0 & 0 & 0 & 0 & 0 & 0 \\ 0 & 0 & 0 & 1 & 0 & 0 & 0 & 0 & 0 & 0 \\ 0 & 0 & 0 & 0 & 1 & 0 & 0 & 0 & 0 & 0 \\ 0 & 0 & 0 & 0 & 0 & 0 & 0 & 0 & 0 & 0 \end{bmatrix}$$

$$z_1 = 5, \quad b_0 = 1$$

We omit all other boundary operator matrices since their domains and/ or codomains equal 0 and hence result in 0-dimensional matrices. It also follows that $z_n = 0$ for all $n \geq 3$ and $n \leq 0$ and $b_m = 0$ for all $m \geq 3$. Recall that Betti numbers yield the ranks of our homology groups and are computed via $\beta_n = |z_n - b_n|$. Hence, we get:

- $\beta_n = |z_n - b_{n+1}| = |0 - 0| = 0$ for all $n \geq 3 \Rightarrow H_n(Y) = 0$ for all $n \geq 3$.
- $\beta_2 = |z_2 - b_3| = |1 - 0| = 1 \Rightarrow H_2(Y) = \mathbb{Z}_2$.
- $\beta_1 = |z_1 - b_2| = |5 - 7| = 2 \Rightarrow H_1(Y) = \mathbb{Z}_2 \oplus \mathbb{Z}_2 = (\mathbb{Z}_2)^2$.
- $\beta_0 = |z_0 - b_1| = |0 - 1| = 1 \Rightarrow H_0(Y) = \mathbb{Z}_2$.

These homology groups now tell us the following information:

- $H_n(Y) = 0$ for all $n \geq 3$: There are no holes of dimension n in Y for $n \geq 3$.
- $H_2(Y) = \mathbb{Z}_2$: There is one hole of dimension 2, that is, a polygonal 2-sphere, in Y.
- $H_1(Y) = \mathbb{Z}_2 \oplus \mathbb{Z}_2 = (\mathbb{Z}_2)^2$: There are two holes of dimension 1, that is, polygonal circles, in Y.
- $H_0(Y) = \mathbb{Z}_2$: There is one hole of dimension 0 in Y, which means Y consists of one connected component.

It should be noted that with homology groups alone, it is impossible to tell *where* the holes occur. This results from the fact that homology computations are *coordinate-free*, meaning that the exact way a simplex has been placed in space and positioned is irrelevant. This is both a huge advantage and disadvantage in studying data. The advantage is that if two people were to look at the same simplex from very different spatial perspectives, their homology computations would agree. For analyzing macro features of a simplex (or data set!) like number of components, this is excellent news. On the other hand, existence information is almost useless if location information is necessary for making actionable observations. As such, homology should be used as a tool in conjunction with others when precise location information is important.

5.3.4 A Continuum: Persistent Homology

Until now, we have defined simplicial homology as a tool for detecting holes in an abstract mathematical space, which we care about since holes in a simplicial complex built from data correspond to clustering phenomena within those data. However, the tool we really want to employ to study data is *persistent homology*, which can be regarded as a continuous version of simplicial homology. Before giving the definition of persistent homology, we will motivate the need for something "more continuous" than what we have seen so far. Recall that for a point cloud with a defined metric, a filtration value r will determine a simplicial complex Δ_r (typically a Čech complex $\mathcal{C}_r$ or a Rips complex $\mathcal{R}_r$), as described in Section 5.1.

Simplicial homology is defined and computed for a single simplicial complex. This means that if we are trying to apply this tool to simplicial complexes created from a data set with a metric, then we can only get homological information (i.e., information about similarity and clustering) about the data set by fixing a single filtration level r and computing the simplicial homology of Δ_r. This is problematic because it is entirely possible that for a specific data set, large-scale clustering does not show up at small filtration values, that more detailed, local phenomena are overlooked by large filtration values, and that both kinds of patterns are obscured by filtration values in between. As such, computing simplicial homology for a single Δ_r could fail to give us helpful information about patterns within our data. Situations like these demonstrate a need to compute the homology of Δ_r for many different values of r for a given data set/metric space.

However, even looking at the homology of data at many different filtration values is not necessarily good enough. In addition to the aforementioned scaling issues, it is also possible that, by only computing homology for a finite number of Δ_r's, two data sets could appear to be the the same when, in fact, they are not. Consider two data sets A and B that are equipped with metrics d_A and d_b, and let Δ_r^A and Δ_r^B be the simplicial complexes for filtration value r for A and B, respectively. Suppose that we compute the homology of Δ_r^A and Δ_r^B for three different filtration levels r_1, r_2, r_3 and we find that the two data sets have identical homology groups for each r_i, $i \in \{1,2,3\}$. It is possible for A to have a hole that appears shortly after r_1 and lasts until just before r_3 while B has a hole that appears shortly before r_2 and disappears shortly after. In this case, the homology groups for A and B at filtration level r_2 are identical and thus fail to capture the fact that the hole in A "persists" for a long time while the hole in B appeared and then disappeared rather quickly. We could remedy this issue by simply computing homology for A and B at many more filtration values (this is what we will do in practice), but ultimately we could experience the same problem with any finite number of filtration values.

As such, persistent homology theoretically computes simplicial homology of a data set for *every* filtration value $r \geq 0$. Since filtration values lie on a continuum (the real line), this means that we compute the homology for a continuum of simplicial complexes Δ_r. Of course, it is physically impossible to record the homology groups for every filtration level $r \geq 0$ along a continuum; therefore, we do the following in practice:

1. Choose some (possibly large) finite number of filtration levels r_i for $1 \leq i \leq k$, compute the corresponding homology groups, and simply acknowledge that some persistence behavior may not be fully captured for the aforementioned reasons.

2. Create *barcodes*, which indicate the number of holes in each dimension $n \geq 0$ and the interval of persistence of each hole over some larger interval of filtration values (usually starting at $r = 0$). Barcodes are created by looking at the Betti numbers β_n at every filtration value. There will be β_n bars in the dimension n barcode, and each bar

will be placed along a number line so that the bar begins at the filtration value at which the corresponding hole appears (the "birth") and ends at the filtration value at which the hole disappears (the "death").

Barcodes are particularly useful for showing the existence of (relatively) significant features in a data set. Short bars correspond to short-lived holes, often formed from noise within the data set; long bars represent holes that persist for long tolerance intervals and therefore represent features that are noise-resistant and, therefore, more significant. In all cases, bars are deemed "short" and "long" relative to one another so that the scale at which we are working is irrelevant.

5.4 Applications for Structural Geology and Tectonics

Patterns of deformation of rocks at the Earth's surface are studied in the geologic fields of structural geology and tectonics (Davis, Reynolds, Kluth, 2012). Observations of structures in rocks are not limited to purely macro-, or outcrop-scale features, such as lithologic contacts (i.e., bedding), faults, fractures/joints, or folds, but also scale down to microscopic features seen in thin-sections of rocks, such as mineral grain orientations (Davis, Reynolds, Kluth, 2012). To understand the structural or tectonic past of a region, geologists must first be able to describe the geometry of structures, which are essentially the architecture of the Earth's crust, and interpret the patterns of deformation present (Karimi and Karimi, 2017). This becomes challenging when faced with multiple events causing overlapping features or with large data sets meant to be interpreted with inevitable human bias. A method/tool for processing structural/tectonic data would be valuable to our understanding of Earth's history in regions, and how it affects the development of crustal material, the emplacement of resources, etc.

The architectural features exhibited by different structures may be planar or linear depending on the type of feature described. Regardless of the feature type, two attributes are necessary—in addition to latitude and longitude—to describe these features in three-dimensional space. The first is orientation with respect to compass directions, which for planar structures is referred to as *strike* and for linear features is referred to as *trend* (Davis, Reynolds, Kluth, 2012). The strike is the trace found at the intersection between the feature plane and a plane horizontal to the surface of the Earth. The trend is the orientation of a line projected onto a plane horizontal to the surface of the Earth. The second attribute describes how the feature is oriented in the vertical direction with respect to the surface, or a horizontal plane. For planar features, we refer to this as *dip*, and for linear features it is called the *plunge* (Davis, Reynolds, Kluth, 2012). Depending on the local geology, there may be many different structures that populate parts or entire areas within a region. The orientation of structures in space with respect to one another allows geologists to gain a deeper understanding of the complexities associated with single or multiple deformation events (Karimi and Karimi, 2017). Analyzing these structures and their patterns ultimately enables geologists to better understand the conditions that allowed for the existing deformation patterns, and it enables them to make stronger predictions as to the genesis of the region.

We present a case study involving two synthetic geologic data sets for analysis to exhibit the strength of our proposed methods and its ability to identify patterns. These data sets are processed to produce barcodes for H_0 and H_1 to identify persistent patterns and the simplicial Rips complex $\mathbb{Z}_2^1$. Visualizations of data at various intervals in the barcode are shown to better explain pattern persistence and connectivity.

5.4.1 Description of Data Sets

The data type we analyze are synthetic vector points representing field measured orientations of structures in rock bodies within a region. To keep the data set simple, we assume all the structures are planar and vertical relative to the surface of the Earth, whose dimensions are: latitude and longitude of the field measurement, and azimuthal orientation of *strike*. Non-vertical dipping structures can also be considered by the tool with the addition of another dimension that would account for the vertical angle of the feature relative to the surface of the Earth. We created two data sets, the first more simple, and the second more complex, to explore the efficacy and power of persistent homology in detecting large-scale topographic features.

The first data set shown in Figure 5.7 is a 4×4 grid of data points with azimuthally oriented strikes of structural features. The corresponding data table is shown in Table 5.1. Latitude (Y) and Longitude (X) are simple cartesian points not based on any geospatial coordinate system; however, latitude and longitude values associated with coordinate systems may be used. There are two azimuthal strike orientations in this data set: 135° located along the northern and eastern perimeters, as well as at location 6, and 45° surrounding location 6 (see Figure 5.7).

Data set 2 (Table 5.2) is visualized in Figure 5.8 and shows a more *natural* distribution of data, both geospatially and in strike orientation. This type of distribution is similar to what we see from authentic structural field data in valley and ridge provinces at orogenic (mountain) belts, and can be thought to represent data such as bedding, joint, fault, or fracture orientations.

Upon visual inspection, there are distinct patterns in both data set 1 (Figure 5.7) and data set 2 (Figure 5.8) that we hope our methods can detect; these patterns are those based on azimuthal similarity. In particular, we hope to pick out the northeast-southwest trending strikes such as the "loop" around

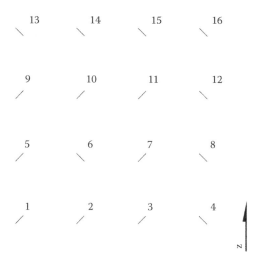

FIGURE 5.7
Data set 1: A 4 × 4 grid of strike-oriented structural data.

observation 6 in Figure 5.7 and the southern swatch of Figure 5.8, as well as the northwest-southeast trending strikes in the respective complements.

For data set 1, the range of X and Y data is from 0 to 3, while the azimuthal data range from 45° to 135°. In data set 2, the cartesian coordinates range from 0 to 9, with azimuths ranging from 30° to 170°. Without normalizing

TABLE 5.1

Data Set 1

ID	X	Y	Azimuth
1	0	0	45
2	1	0	45
3	2	0	45
4	3	0	135
5	0	1	45
6	1	1	135
7	2	1	45
8	3	1	135
9	0	2	45
10	1	2	45
11	2	2	45
12	3	2	135
13	0	3	135
14	1	3	135
15	2	3	135
16	3	3	135

TABLE 5.2

Data Set 2

ID	X	Y	Azimuth
1	0	0	45
2	1	0.5	168
3	2	5	162
4	3	1	158
5	0	1.2	40
6	1	2.5	170
7	2	7	165
8	3	9	160
9	0.5	4	35
10	1.8	2	50
11	2.2	2	65
12	2.9	2	80
13	0.2	3	30
14	1	3	44
15	2	3	59
16	4	3	74

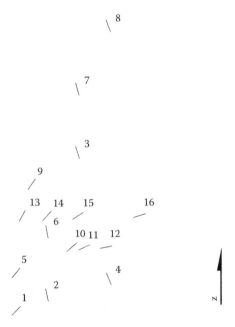

FIGURE 5.8
Data set 2: A more realistic geospatial representation of structural features oriented according to their strike.

each dimension, pattern connections are expected to be determined mostly by azimuthal similarities more so than geographic coordinates. When these methods are used for arbitrary data sets, considerations should be made as to what the most impactful dimensions should be, or if all dimensions are (or should be) weighted equally.

5.4.2 Data set Processing

To use our method of persistent homology, the first step is to identify the dimensions/variables of our observations upon which we wish to define the requisite *distance function d*. In our examples, we use the latitude (X), longitude (Y), and azimuthal strike orientation (θ). We then define the distance d between two points $a = (X_1, Y_1, \theta_1)$ and $b = (X_2, Y_2, \theta_2)$ by

$$d(a,b) = \sqrt{(X_2 - X_1)^2 + (Y_2 - Y_1)^2 + \min(|\theta_2 - \theta_1|, 180 - |\theta_2 - \theta_1|)^2}.$$

Since our azimuths are all within the first 180°, we do not need to make adjustments to the data to have them fall within the same eastern hemisphere; however, we do need to stay consistent and to that end only consider the smallest angular difference between the azimuthal strike of two lines.

For each data set (Tables 5.2 and 5.3), we started with an excel spreadsheet with columns for latitude, longitude, and azimuthal strike for each of our datapoints. We then wrote the python script to calculate the distance between all pairs of points to generate a matrix, such as the one for data set 1 in Table 5.3. From the data in Table 5.3, we can see that our distance function satisfies the four requirements: nonnegativity, definiteness, symmetry, and the triangle inequality. We then process the matrix with an algorithm to create barcodes showing the Betti numbers of Rips complexes, built from our data set and distance function, at a continuum of filtration values. The barcode distinguishes homology groups based on holes of dimension n, and the number of holes in each dimension $n \geq 0$ and their persistence over the continuum of filtration values, or intervals. From here, we developed a tool in ArcMAP, a geospatial software created by ESRI®, to visualize the connectedness among patterns associated with user chosen points (specific filtration values) along the barcode. The tool adheres to the following steps:

1. Accept a user-defined interval (t) along the barcode at which to make a visual. These intervals can essentially be thought of as similar to *time*. If we think about our 3-dimensional data set in 3-dimensional cartesian space, the interval relates to the size of a growing sphere of influence around each point. When two spheres touch, their respective points of origin are considered to be connected.

2. Create an empty polyline shapefile and add a field, *"pattern."* This field gives a unique identifier to connected patterns.

TABLE 5.3

Distance Matrix for Data Set 1

ID	1	2	3	4	5	6	7	8	9	10	11	12	13	14	15	16
1	0.0000	0.5000	1.0000	45.0250	0.5000	45.0056	1.1180	45.0278	1.0000	1.1180	1.4142	45.0361	45.0250	45.0278	45.0361	45.0500
2	0.5000	0.0000	0.5000	45.0111	0.7071	45.0028	0.7071	45.0139	1.1180	1.0000	1.1180	45.0222	45.0278	45.0250	45.0278	45.0361
3	1.0000	0.5000	0.0000	45.0028	1.1180	45.0056	0.5000	45.0056	1.4142	1.1180	1.0000	45.0139	45.0361	45.0278	45.0250	45.0278
4	45.0250	45.0111	45.0028	0.0000	45.0278	1.1180	45.0056	0.5000	45.0361	45.0222	45.0139	1.0000	2.1213	1.8028	1.5811	1.5000
5	0.5000	0.7071	1.1180	45.0278	0.0000	45.0028	1.0000	45.0250	0.5000	0.7071	1.1180	45.0278	45.0111	45.0139	45.0222	45.0361
6	45.0056	45.0028	45.0056	1.1180	45.0028	0.0000	45.0028	1.0000	45.0056	45.0028	45.0056	1.1180	1.1180	1.0000	1.1180	1.4142
7	1.1180	0.7071	0.5000	45.0056	1.0000	45.0028	0.0000	45.0028	1.1180	0.7071	0.5000	45.0250	45.0222	45.0139	45.0111	45.0139
8	45.0278	45.0139	45.0056	0.5000	45.0250	1.0000	45.0028	0.0000	45.0278	45.0139	45.0056	0.5000	1.8028	1.4142	1.1180	1.0000
9	1.0000	1.1180	1.4142	45.0361	0.5000	45.0056	1.1180	45.0278	0.0000	0.5000	1.0000	45.0250	45.0028	45.0056	45.0139	45.0278
10	1.1180	1.0000	1.1180	45.0222	0.7071	45.0028	0.7071	45.0139	0.5000	0.0000	0.5000	45.0111	45.0056	45.0028	45.0056	45.0139
11	1.4142	1.1180	1.0000	45.0139	1.1180	45.0056	0.5000	45.0056	1.0000	0.5000	0.0000	45.0139	45.0056	45.0056	45.0028	45.0056
12	45.0361	45.0222	45.0139	1.0000	45.0278	1.1180	45.0250	0.5000	45.0250	45.0111	45.0139	0.0000	1.5811	1.1180	0.7071	0.5000
13	45.0250	45.0278	45.0361	2.1213	45.0111	1.1180	45.0222	1.8028	45.0028	45.0056	45.0056	1.5811	0.0000	0.5000	1.0000	1.5000
14	45.0278	45.0250	45.0278	1.8028	45.0139	1.0000	45.0139	1.4142	45.0056	45.0028	45.0056	1.1180	0.5000	0.0000	0.5000	1.0000
15	45.0361	45.0278	45.0250	1.5811	45.0222	1.1180	45.0111	1.1180	45.0139	45.0056	45.0028	0.7071	1.0000	0.5000	0.0000	0.5000
16	45.0500	45.0361	45.0278	1.5000	45.0361	1.4142	45.0139	1.0000	45.0278	45.0139	45.0056	0.5000	1.5000	1.0000	0.5000	0.0000

3. Divide the distance values in the matrix by 2. In our symmetrical matrix, since $d(a,b) = d(a,b)/2 + d(b,a)/2$, when two points' $d/2 < t$, they are not considered to be part of the pattern at that interval value. If $d/2 \geq t$, then the two spheres of influence touch and are considered part of a pattern.

4. Since $d(a,b) = d(b,a)$ and $d(a,a) = 0$, remove data from the matrix below, and including, the diagonal. This reduces the number of data to be processed as well as the processing time.

5. Identify all data point pairs where $d/2 \geq t$.

6. Select any one of the identified pairs and connect their points with a line, entering a value of 1 in the *pattern* field.

7. Identify any other point that is connected to either of the two points in step 6, and draw a line between them, again entering a value of 1 in the *pattern* field.

8. Identify any other point that are connected to those identified in step 6 and continue drawing lines with *pattern* values of 1. Continue this until no points can be connected back to any of the points in pattern 1.

9. Remove the points associated with pattern 1 from the selected items in step 4.

10. Repeat steps 6 through 9, each time increasing the pattern value by 1, until there are no more lines point pairs available.

These steps provide a linear shapefile that can be classified by pattern value to show the connectedness of data among patterns. The creation of so many lines can be computationally expensive, even though the final product is visually appealing and useful in its details for understanding how the patterns evolve over time. An alternative, computationally less expensive protocol would create a unique pattern identifier among the original data set, rather than drawing lines. To better describe the mathematical approach and results, and since our data set is small, we chose to create the linear shapefile.

5.4.3 Results and Discussion

5.4.3.1 Data Set 1

The barcode and examples of patterns at different intervals for data set 1 can be found in Figure 5.9. From the barcode for this data set for H_0, each individual data point persists as a pattern until an interval of 0.5 (Figure 5.9a). At an interval of 0.5, three new patterns emerge from connections made between data points (Figure 5.9b): the pattern in blue, the pattern in red, and data point 6. In Figure 5.9c, at an interval of 1.0, data point 6 merges with the red pattern group. The red and blue patterns persist until an interval value of 45.056. At this value, all both patterns merge into a single pattern.

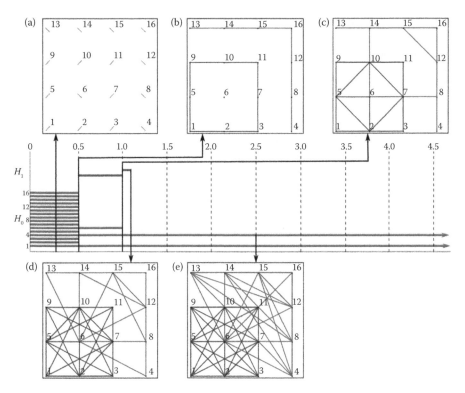

FIGURE 5.9
Barcode for structural features in data set 1. Visualizations for intervals: (a) 0.28, (b) 0.5, (c) 1.0, (d) 1.118, and (e) 2.5.

At a rank of H_0 many barcode items are identified, and two patterns are distinguished at an early stage, remaining persistent for a large interval range. The barcode associated with H_1, however, identifies something rather unique about how the data are clustered. At different interval ranges, there are two hollows identified using simplicial Rips complexes. At an interval of 0.5, the first $\mathbb{Z}_2^1$ hollow is created (blue pattern in Figure 5.9) and persists until an interval value of 1.0. From 1.0 until 1.118, another $\mathbb{Z}_2^1$ hollow persists. This hollow is in the red pattern group, and is a polygon identified by points 6, 8, 12, 15, and 14. These hollows are rather short-lived given their interval ranges, but had they been persistent for long interval ranges, they may have had a significant impact on the interpretation of the geologic history in our synthetic region.

The prolonged persistence of the two patterns in H_0 is by far the most important aspect for a geologist who is seeking distinct patterns that may be related to natural forces in data set 1. Our methodology and tool are very effective in this instance; however, this data set is an idealized 4×4 model. We must explore more complex, seemingly "natural" data sets.

5.4.3.2 Data Set 2

Data set 2 is a more natural distribution of geospatial, oriented structural data, modeled after structural data from orogenic belts. The resulting barcode for this data set can be found in Figure 5.10. Unlike the barcode for data set 1, there were no ranks higher than H_0. Until an interval value of 1.414, the 16 data points are recognized as 16 different pattern sets (Figure 5.10a). As the interval value increases from this point, connections are made, and at an interval of 2.872 only 6 pattern sets exist (Figure 5.10b): the blue pattern set, the red pattern set (3, 4, 7, and 8), points 2 and 6, point 11, point 12, point 15, and point 16. It is not until an interval value of 4.637 that only two pattern sets remain, and these sets persist until 20.006. Figure 5.10c represents this pattern set at an interval of 5, although this pattern persists much longer. At 20.006, connections between the red and blue pattern sets start to form, identifying a single pattern set persistent at all higher interval values (see Figure 5.10d).

The distinct patterns that we can visually assess for data set 2 (discussed in Section 4.1), are persistent for a fairly long range (4.637–20.006). These long, persistent interval ranges represent interval values that most likely best highlight pattern sets. However, a strength of our methods is the opportunity to explore how soon/late pattern sets are considered merged into larger pattern sets, allowing for a stronger analysis of similarities between patterns. Geologically, this becomes rather important, particularly with knowledge regarding how differently oriented structures are related to a single

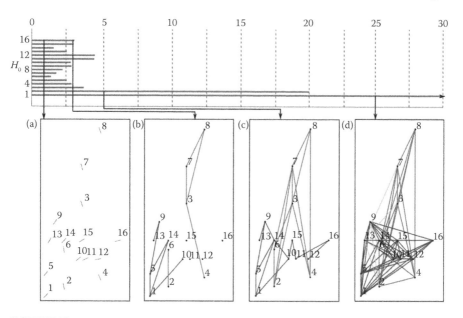

FIGURE 5.10
Barcode for structural features in data set 2.

stress field orientation. Seeing how these patterns interact with one another as interval values increase may allow for more detailed interpretations and conclusions regarding the geologic history of a region.

5.5 Concluding Remarks

The goal of this chapter was to describe a mathematical tool that can be used to recognize patterns of similarity in geospatial data. Our described method and short case study of synthetic geologic structural data types have accomplished this goal. The visualization of patterns at different intervals provides an extremely powerful tool for researchers to effectively analyze patterns, how they connect, and when they connect. While this chapter uses a structural geology and tectonic context for its example, it is important to note that this tool could be effective for any geospatial data within any discipline.

There is much left to explore still, such as data types: within the realm of geology, there are qualitative data sets—such as rock type—that with unique approaches could be quantified and considered in similarity patterns. Additionally, further research must include the effects of normalizing each dimension, increasing the number of dimensions considered by the distance function, and more efficient computational adaptations of persistent homology. This method will require much adaptation and exploration to fully understand its limits, but the benefit of such methods/tools in recognizing patterns is clear.

References

Davis, G. H., Reynolds S. J., and Kluth C. F. 2012. *Structural Geology of Rocks and Regions*. 3rd edition. New York, USA: John Wiley & Sons, Inc.

Edelsbrunner, H., and Harer J. L. 2010. An introduction. In: *Computational Topology*. Providence, RI: American Mathematical Society.

Gallian, J. A. 2010. *Contemporary Abstract Algebra*. 7th edition. Belmont, CA: Brooks/ Cole, Cengage Learning.

Guillemin, V., and Pollack A. 2010. *Differential Topology*. Providence, RI: AMS Chelsea Publishing, Reprint of the 1974 original.

Jäger, G. 2003. *Parallel Algorithms for Computing the Smith Normal Form of Large Matrices*. Berlin: Springer, 170–179.

Karimi, B. and Karimi H. A. 2017 February. An automated method for the detection of topographic patterns at tectonic boundaries. In *PATTERNS 2017: The Ninth International Conference on Pervasive Patterns and Application*, edited by. Y. Manaert, Iwahori Y., A., Mirnig, A., Ortis, C., Perez, and J., Daykin, 72–77. Athens, Greece: ThinkMind.

Kumaresan, S. 2005. *Topology of Metric Spaces*. New Delhi: Narosa Publishing House.

Munkres, J. R. 1975. *Topology: A First Course*. Englewood Cliffs, NJ: Prentice-Hall, Inc.

Stewart, I. 1995. *Concepts of Modern Mathematics*. Mineola, NY: Dover Publications.

6

LiDAR for Marine Applications

Ajoy Kumar and Nathan Murry

CONTENTS

6.1 Introduction

Light Detection and Ranging, or LiDAR (also known as laser altimetry), is an optical remote sensing technology for generating precise and directly geo-referenced spatial information about the shape and surface characteristics of the Earth (NOAA 2012). It has become a primary means for collecting very dense and accurate elevation data across landscapes, shallow-water areas, and project sites. LiDAR was initially developed during the 1960s by the National Aeronautics and Space Administration (NASA) to better measure properties of the earth such as ice sheets, the ozone layer, and atmospheric pollutant levels. At first, LiDAR was not designed or used for topographical mapping, primarily due to a lack of a strong network for geo-referencing. However, by the 1990s, an expanded GPS network was established, making extended topographic expeditions worthwhile. This resulted in large volumes of LiDAR data becoming available to the public and the scientific community. While NASA continues to be at the forefront of LiDAR usage and development, numerous private enterprises and foreign governments have become increasingly involved in further development of LiDAR technology for advanced GIS-based analysis. Despite the fact that there are a number of established commercial and governmental websites that provide access to the extensive and ever-growing compendium of LiDAR data, the processing methods and applications are largely developed by the end users of the data itself. Advancements in LiDAR technologies over the past 10–15 years have enabled scientists and geospatial professionals to examine natural and man-made environments across a wide range of scales with great accuracy, precision, and flexibility.

From a technical perspective, LiDAR is an active remote sensing technique similar to radar but uses laser light pulses instead of microwaves (NOAA 2012). These light pulses are intense, focused beams of light which are emitted and then reflected off any of the various surfaces on the ground. A sensor onboard the collection platform detects when these light emissions are returned. Geospatial three-dimensional (3D) coordinates (such as x, y, z or latitude, longitude, and elevation) of the target objects are computed from the time delta between the laser pulse being emitted and returned, the angle at which the pulse was sent, and the GPS location of the sensor on or above the surface of the Earth. LiDAR systems are considered "active" remote sensing systems, as they emit pulses of light and then detect the reflected pulses. This fundamental characteristic allows data to be collected at night when the air is typically clearer and the sky less congested with air traffic. Consequently, most LiDAR data collection missions are flown at night. The drawback to light pulses is that they cannot penetrate clouds, rain, or haze, which is a notable difference between LiDAR and RADAR, the latter of which can penetrate these natural obstacles with microwave pulses. While data collection missions are usually done from the air, where a system can rapidly collect points over large areas, many are also run from ground-based stationary and mobile platforms. All of these techniques are popular within commercial and scientific communities due to their advanced capabilities in producing extremely high accuracies and point densities, thus permitting the development of precise, realistic 3D models of most man-made or natural structures on the earth's surface (Table 6.1).

LiDAR systems can rapidly measure the Earth's surface, at sampling rates greater than 150 kHz, or 150,000 pulses per second (NOAA 2012). The resulting product is a densely spaced network of highly accurate georeferenced elevation points, known as a point cloud. Point clouds are used by GIS or other analysis software packages to create 3D representations of the Earth's surface and its features. Many LiDAR systems operate in the near-infrared region of the electromagnetic spectrum, although some sensors also operate in the green band to penetrate water and detect bottom features. These bathymetric LiDAR systems can be used in areas with relatively clear water to measure seafloor elevations. Typically, LiDAR-derived elevations have absolute accuracies of roughly 4–8 inches or 10–20 centimeters (NOAA 2012). To arrive at this level of accuracy, it is important to know within a centimeter or so where the data collection platform is located spatially as it travels. This is particularly important in an airborne platform as it flies, potentially through turbulence, while keeping track of hundreds of thousands of LiDAR pulses every second. Fortunately, advancements in global positioning systems (GPS) technologies and precision gyroscopes make it possible.

TABLE 6.1

LiDAR Data Classifications (NOAA)

Classification Value	Meaning
0	Never classified/default
1	Unassigned
2	Ground
3	Low vegetation
4	Medium vegetation
5	High vegetation
6	Building
7	Low point
8	Reserved
9	Water
10	Rail
11	Road surface
12	Reserved
13	Wire—guard (shield)
14	Wire—conductor (phase)
15	Transmission tower
16	Wire-structure connector (insulator)
17	Bridge deck
18	High noise
19–63	Reserved
64–255	User definable

Additionally, significant improvements in inertial measuring units (IMU) or inertial navigation systems (INS) have been vital to making the exact positioning of the platform possible. These systems are capable of measuring movement in all directions and parsing these measurements into a position. They are not perfect or foolproof, and lose precision after only a second or so. A sophisticated GPS unit, which records several types of signals from GPS satellites, is used to update the INS and IMU every second or so. The GPS positions are recorded by the plane and also at a ground station with a known position. The ground station supplies a correction value GPS position recorded by the plane.

6.2 LiDAR Data Acquisition and Initial Processing

The data used in this study were obtained using NASA's Airborne Topographic Mapper (ATM), an advanced LiDAR system capable of multiple-return data acquisition. LiDAR data collection missions were flown over the Barrier Islands along the Eastern Shore of Delmarva Peninsula (Krabill et al. 2000). The flight tracks, shown in Figure 6.1, cover most of Assateague Island and Wallops Islands, VA. The Nature Conservancy (TNC) provided a Digital Terrain Models (DTMs) of the same area created using data from a previous mission, which proved valuable for filling in the gaps in the LiDAR coverage acquired on this ATM flight.

Prior to analysis in GIS software, LiDAR data must undergo a series of pre-processing steps as shown in Figure 6.2. This process chain converts the data into an accessible file format and transforms it to the local coordinate system desired for the end products. For this study, ATM provided two types of proprietary binary data, QFIT and VALID. The VALID file format provides

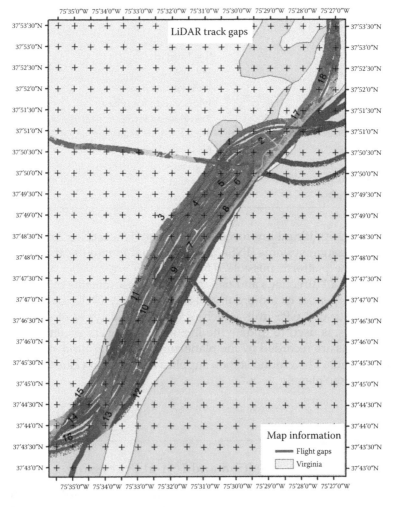

FIGURE 6.1
Flight tracks and gaps in the April 2010 ATM LiDAR data acquisition flight.

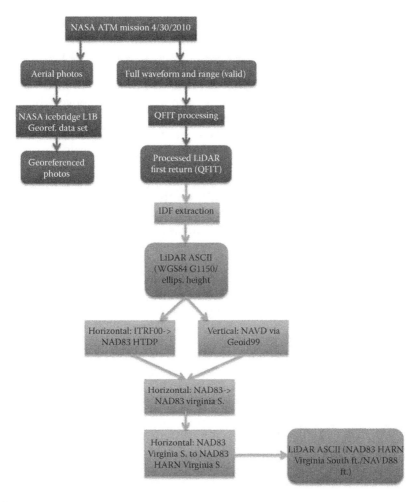

FIGURE 6.2
LiDAR data preprocessing chain.

full waveform data that encompass all the returns detected by the sensor. This affords more control to the end user in the interpretation process of the physical measurement. QFIT data contains one return only. In the case of a more focused study, this data format is more desirable, particularly in the case of the system's concise first return data. More time can be dedicated to the processing and validation of a smaller dataset, allowing a higher level of refinement and a better final product. Both data types were originally formatted using the QFIT 12-word method, which is an organization of (12) 32-bit binary words, equivalent to an Interactive Data Language (IDL) long integer. These words are scaled appropriately in order to maintain the precision of the original measurements (scaling factors are standard for the qi-12-word format). The binary words contain a series of 12 element records corresponding to the individual LiDAR returns, which contain geospatial position of the return, time stamp, returned energy, and other attributes.

QFIT binary data must first be converted to ASCII text format in order for it to be properly analyzed, rendered, and displayed. This first conversion was accomplished using IDL software via custom written scripts. Included in these scripts are the scaling factors mentioned above, which convert the raw data to numeric values (latitude, elevation, GPS time, etc.). The script allows the user to choose which data words to convert and extract, as opposed to being made to convert all 12 (for larger datasets, this feature noticeably cuts down the time needed for conversion). For the purposes of this study, only longitude, latitude, elevation, and received energy strength was processed and converted to ASCII (in that order, which is critical for scripts and rendering down the processing chain). Additionally, the script is capable of cropping the data to specific areas (rather than an entire dataset), and processing the data to different geographic coordinate systems based on the user's requirements. Script conversions included UTM/WGS84, XYZ/ geographic LAT/LON, and varying forms of decimal degrees. This allows multiple avenues of analysis and later comparison of results. For this study, data were processed in UTM and geographic LAT/LON. Since these scripts were critical to downstream processing, considerable time was spent testing and validating the output against known results, as it is essentially the base of the processing chain for non-GIS analysis (Note: Validation methods are covered more in depth in the following section). In this case, the WGS84 Ellipsoid in the International Terrestrial Reference Frame (ITRF) was used at the time of the survey, which caused all data to be recorded in that coordinate system from the sensor. Unlike traditional geographic coordinate systems, the ITRF is a dynamic reference frame that is updated to improve its accuracy and reflect the changing geographic status of the planet, as in the case of moving tectonic plates, by means of a large network of ground control stations. The WGS84 coordinate system is improved through the use of ITRF by periodically realigning it to the most current iteration of the ITRF at the time of processing. For this study, the G1150 realization of WGS84 was used, which aligns to the ITRF00 frame introduced in 2002. When executed, the IDL script extracted the aforementioned QFIT data elements in the WGS84 G1150 system as well as the Universal Transverse Mercator (UTM) coordinate system and XYZ/geographic LAT/LON for comparison analysis.

Evaluation of the impact of sea level rise (SLR) and flood inundation requires that the data be presented in context to the region around Wallops Island, VA. This necessitates conversion from the data's native WGS84 G1150 coordinate system, which is globally relative, to the North American Datum 1983 (NAD83) which adjusts the representative ellipsoid to maximize congruence with the North American continent. This can be accomplished by using the defined 14-parameter horizontally time-dependent positioning (HTDP) ITRF00 to NAD83 transformation (with a time parameter of April 30, 2010), which consists of time factored scaling, translation, and rotation parameters. After the conversion to NAD83, a horizontal conversion is made to the NAD83 State Plane Virginia South coordinate system, and finally to the NAD83 HARN State Plane Virginia South coordinate system. This represents a further refinement of the NAD83 datum to ensure that the southern area of Virginia is most consistent with the ellipsoid. In regard to vertical heights, the original measurements are made relative to the ellipsoidal height of the WGS84 ellipsoid; these measurements are converted to the North American Vertical Datum 1988 (NAVD88) by using the Geoid99 model which consists of a grid of calculated displacements. The above geographic conversions are performed using a Geographic Calculator, a geospatial transformation toolbox from Blue Marble Geographics, Inc. This software tool provides batch data file processing, flexible ASCII input and output formatting, and support for HTDP, among other linear unit transforms using an extensive database of geographic coordinate systems and transformations. It must be noted that when initially rendered in any of the available tools, the terrain elevations all exhibited an offset of approximately –37.5 m. This is due to a disparity between the data gathered in UTM (ellipsoidal heights), the geoid of the earth, and the reference datum. Initially, LiDAR data were collected in UTM projection, with a point density of one per square meter. UTM is referred to as "pseudocylindrical" by cartographers and geographers since it is designed to preserve the perceived shape of the Earth's surface. Thus emerges a disparity between the ellipsoid, geoid, and reference datum of the Earth, which clearly presents an issue for researches of the Mid Atlantic coastal environment. Furthermore this explains the unusual negative elevation values on the LiDAR data, as the UTM projection (ellipsoidal) was below the geoid of the earth. To meet the end goal of the project elevation values, they had to be changed to sea level values to model effects of sea level rise and flood inundation from storms (Webster et al. 2003). Thus, a transformation of datum height values was required to attain the proper elevation readings, which necessitated a conversion of the data in both vertical and horizontal planes. This conversion was accomplished using a Geographic Calculator in three distinct steps. The first step was to take the data, which were formatted in WGS84 coordinates, and convert them to NAD83/VA SPCC via the Blue Marble ITRF00 to NAD83 HTDP transform (dated 4/30/10). The second step was to convert the data, now formatted in NAD83/VA SPCC, to NAD83 HARN SPC VA South via the Blue Marble NAD83 to NAD83 HARN transform. The all-important third step was to convert the vertical plane from WGS84 Ellipsoidal Heights to the North America Vertical Datum 1988 using the Blue Marble GEIOD99 grid. Once all of these conversions are complete, the final ASCII files are ready for the next stage of processing.

While providing human readability and a degree of flexibility, ASCII files are inefficient as a means for storing and manipulating LiDAR data. Instead, the LiDAR points are converted (through differing means) to LiDAR Archive Standard (*.LAS) files, which is the industry standard file format designed specifically for the containment of LiDAR specific information for exchange or dissemination. It is a binary compact format (as compared to ASCII), and easy to read and use within a wide range of GIS, CAD, and other LiDAR data processing tools. LAS files can be analyzed individually (given their potentially enormous file sizes) or collectively. For more of a macro data analysis (as this study was), it proved more beneficial to merge the LAS files together for a more comprehensive analysis. In this study, conversion and merging of ASCII files to LAS was done using ENVI's LiDAR toolkit. These operations were also test-run in LASTOOLS, an open-source software package for processing LiDAR data. Processing in LASTOOLS allowed a comparison between the two LiDAR outputs. The LAS data now move to final product generation and validation.

6.3 Generation and Validation of Final Products

In order to generate DTMs that accurately represent bare earth surfaces or terrain, a process of classification must be undertaken in which valid data points are "classified" as various potential ground features such as vegetation, water, and so on. This process also identifies the "outlier" or invalid data points and marks them for removal in subsequent processing stages. Such invalid points can result from pulse returns off birds, aircraft, clouds, ocean waves or mist, and so on. The overall flow of classification is detailed in Figure 6.3. Two commercial software packages were used extensively for this series of steps: Blue Marble Geographics Global Mapper and TerraSolid. Global Mapper software has features that proved extremely useful as a rendering and profiling tool for data formatted elsewhere. TerraSolid is a state-of-the-art modular CAD system for LiDAR data processing and imaging, run on a powerful workstation to take maximum advantage of its features. In particular, the TerraScan module was used for loading the LAS cloud-point files, filtering out artifacts, and creating the DTM cloud-point by eliminating buildings, trees, and other non-terrain elements (the DTM cloud-point data could then be used in Global Mapper for detailed color renderings and beach profiles). The TerraModeler module was used to create various surface models from the cloud-point data, colored by absolute elevation. While these models are excellent for scientific analysis, similar renderings created in Global Mapper are easier to manipulate and add pertinent data for presentation purposes. It should be noted that extensive testing and experimentation went into determining the best way to display the data (in TerraScan or Global Mapper), how to remove terrain features (be they man-made or otherwise), render the DTM, and prepare the results for presentation. Once the LAS cloud-point data were read into TerraScan, it could be filtered and modeled as needed using tools from TerraSolid and Microstation. Since this started as QFIT rather than VALID format, there is only one return to work within the converted LAS files. If the

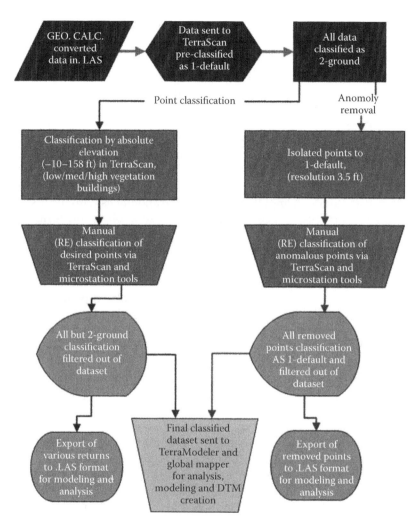

FIGURE 6.3
LiDAR data classification and modelling process chain.

data are colored by return, obviously only one color will display, which is less than helpful. This can be initially dealt with by changing the display mode to color the data by elevation, which can give the user a better context of the data. From the onset of loading data in TerraScan, all data points are classified as "1-Default." Using the Classify > Routine > By Class algorithm, all points were reclassified as "2-Ground." This provided a base class where all remaining terrain points would stay, while filtered points would be classified elsewhere. Note from here that the entire methodology of creating DTMs in TerraScan and Global Mapper is based on proper classification of points, and then modeling or exporting those classes that are desired, while eliminating those that are not. Now that all points are classified as "2-Ground," filtering can begin. There were two areas that were processed heavily: the first is the southern peninsula of Assawoman Island, VA; the second was the entire NASA Flight Facility at Wallops Island, VA. The base has numerous buildings, towers, and other man-made structures that needed to be removed for the creation of a DTM, which made data classification a more extensive undertaking. Conversely, the Assawoman peninsula has no such features and its Digital Surface Models (DSM) is essentially a DTM as it requires only some minimal filtering. It is worth noting that the dataset as a whole was littered with various artifacts and anomalies, which rendered as tall spikes in the models and caused the elevation coloring to be skewed considerably due to the software tool's attempt to triangulate the anomalies with the rest of the dataset. These artifacts can generally be explained as blips during/after data acquisition, flocks of birds, wave spray, and so on. The first step in filtering was to use the Classify > Routine > Isolated Points algorithm in TerraScan to a resolution of 3.5 feet. In other words, the algorithm examines all the points in the loaded set. If it comes across a point in the data space that has no other points within 3.5 feet of itself, it is classified as an isolated point and placed in the "1-Default" class. The Default class would be the "trash-can" of sorts for anomalies and other such useless points. In the case of the Wallops base, the next step was to classify points by the Classify > Routine > By Absolute Elevation algorithm in TerraScan. This made it possible to break the overall elevation range (approximately –10 to 158 feet) into pieces and classify data points by low, medium, or high vegetation, and buildings. Once the building points and high vegetation were filtered out, the rest of the points were regrouped into one classification ("2-Ground") for modeling in TerraModeler and Global Mapper. Extensive trial and error for elevation parameters was tested here, as the balance had to be achieved between filtering out what was needed (buildings, anomalies, etc.), while not scraping into the "bare earth," and eliminating good data. While the algorithms detailed above provided about 90% of the needed filtering, it was still necessary to perform some of classification manually. This was accomplished by using the Microstation "View Rotation" and "Pan" tools to view the data from the right sides (perpendicular to the ground), zoom in to the southernmost point of the data, and steadily scan northward. When anomalous or non-terrain points are encountered, they can be corralled by the Microstation "Fence" tool, and then classified by the Classify > By Class tool, selecting the appropriate class (being sure to check "Inside Fence," which applies the classification only to the fenced in points). This manual technique for classification was very effective for micro-filtering, but also very tedious and time consuming (this is a good example of why the cropping/filtering of data at the ENVI Script stage was so important). Once all of the classification was completed to an acceptable level, the cloud-point data were re-saved to their own LAS file. This allowed a proper comparison to the unclassified and unfiltered data in subsequent processing steps. At this point, the filtered and classified cloud-point data were ready to be modeled in TerraModeler and rendered in Global Mapper. TerraModeler takes the cloud-point data and creates a 3D surface model from them, exquisitely detailed and colored by absolute elevation. Rendering, profiling, and additional 3D images of the data were then performed in Global Mapper. These operations did not modify the data, but only changed parameters that dealt with the map legends, scales, and coloring. Beach/land profiles were obtained with Global Mapper's Profiling Tool, with comparisons made between the unfiltered DSM, the filtered/classified DTM, and VA State data from the exact same latitude/longitude positions. These images are found and explained in more detail in the Results section.

To evaluate the performance of the classification exercises above, the bare earth model (DTM) needed to be validated against known ground truth points, which was done in two stages. First, the TNC LiDAR data were validated against 359 control points provided by NASA. Before calculation of the root mean square (RMS) error, the outliers and residuals greater than +/–1 foot were removed. There were 21 such outlying points found and eliminated when comparing the TNC DTM to the NASA control points. The RMS error value for these remaining 338 points was 0.430 feet, or 13.113 centimeters. This was then considered the approximate "goal" of accuracy. In the next step, the DTMs processed in this study were validated with the TNC dataset and a reduced set of NASA control points (105). This reduced set was necessitated due to gaps in the processed DTM shown in Figure 6.1. Hence, once the final DTMs were created, all 359 control points were loaded in TerraScan. An Output Control Report was run, and any control points outside the processed data range were eliminated. This new set of control points was then compared to the same DTM, only rendered in Global Mapper. A similar report was run and points outside the processed data range were eliminated. (NOTE: there are slight differences in how TerraSolid and Global Mapper render datasets, which is why there were some control points that were valid in one program and not the other, and vice versa). There were three primary RMS error analyses performed against these 105 control points: the Method 2 DTM in Global Mapper, the Method 2 DTM in TerraScan, and the TNC DTM in Global Mapper. The DTMs outlined in this study are roughly 2 cm better than the TNC DTM against the same 105 control points, and extremely close to the TNC DTM baseline. The residual comparisons between the custom data processing chains in Global Mapper and Terrasolid against the TNC processing chains are shown in Tables 6.2 and 6.3.

6.4 Results and Discussion

Figure 6.4 shows the DEM created with LiDAR data obtained from the survey conducted with the NASA ATM sensor. The figure provides a good

TABLE 6.2

RMS Comparisons between Custom Processing in Global Mapper and TNC
Processing

	Method 2 DTM Residuals (105 ctrl pts.), Global Mapper		TNC DTM Residuals (105 ctrl pts.) Global Mapper	
	FT	CM	FT	CM
AVG DZ	−0.057428571	−1.750422857	−0.234457143	−7.146253714
MIN DZ	−0.866	−26.39568	−1.329	−40.50792
MAX DZ	0.945	28.8036	0.871	26.54808
RMS	0.437931393	13.34814885	0.49454125	15.07361729
STD DEV	0.436231851	13.29634683	0.437520264	13.33561764

visual indication of the coverage density resulting from the ATM scan. The
region covers parts of the NASA Wallops Flight Facility (WFF) and the surrounding salt marshes. The higher elevations, represented in red, are buildings located at the NASA WFF. The elevations in dark green depict roads
and bridges. Note also the causeway at the entrance to the island in the NW
corner of the image. Elevations in light blue and light green portray channels
and salt marshes, respectively. The ocean waters are depicted in dark blue
and are not processed well in this study due to low laser backscatter from
water surfaces. Overall, the LiDAR-derived DEM portray ground features
illustrated by elevation.

Figure 6.5 shows the bare earth model or DTM of the same area after
removal of buildings, trees, and other structures by the classification method
described above. The highest elevation in DTM is ~5 feet compared to the
DEM of nearly ~30 feet. The LiDAR-derived terrain surfaces can be rendered
in models to illustrate low-lying areas that would flood especially during
winter storms and SLR. The success of the DTM creation can be seen in
Figure 6.6, which shows what was eliminated from the DEM to arrive at the
DTM. It also illustrates that the classification works best on the NASA base

TABLE 6.3

RMS Comparisons between Custom Processing in TerraScan and TNC
Processing

	Method 2 DTM Residuals (105 ctrl pts.), Terrascan		TNC DTM Residuals (105 ctrl pts.) Global Mapper	
	FT	CM	FT	CM
AVG DZ	−0.053714286	−1.637211429	−0.234457143	−7.146253714
MIN DZ	−0.871	−26.54808	−1.329	−40.50792
MAX DZ	0.945	28.8036	0.871	26.54808
RMS	0.442950583	13.50113377	0.49454125	15.07361729
STD DEV	0.441790494	13.46577426	0.437520264	13.33561764

FIGURE 6.4
Initial DEM final product from Global Mapper (minus outlier points).

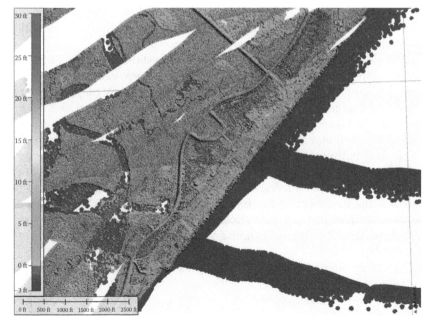

FIGURE 6.5
DTM final product from Global Mapper, depicting the apparent bare earth terrain.

FIGURE 6.6
A model of the feature points that were removed during DTM creation.

where there are high-rise structures. The salt marsh regions are essentially bare earth and do not show up in the difference image.

LiDAR-derived data can be used for many coastal applications including studies involving shoreline changes and mapping, episodic erosion, coastal geomorphology, land use planning, coastal inundation mapping, habitat analysis, and vegetation classification (https://coast.noaa.gov). Some applications derived from this study are described below.

LiDAR-derived elevation data are used to produce high-resolution topographic and bathymetric maps over shallow regions. These maps can be used for extracting shoreline positions and quantifying shoreline changes. Figures 6.7 and 6.8 are profiles taken from the DEM and DTM. These profiles track the elevation of the bare earth and beaches in this location. If validated against spot elevations determined with the "total station" survey instrument, they can be of immense value to study erosion and accretion processes.

LiDAR technology can play an important part in environmental conservation and restoration.

Many coastal species of plants and animals use small elevation changes to build their habitat.

An illustration of LiDAR habitat application is shown in Figure 6.9. The brown areas in the image depict beach elevations on the South side of the

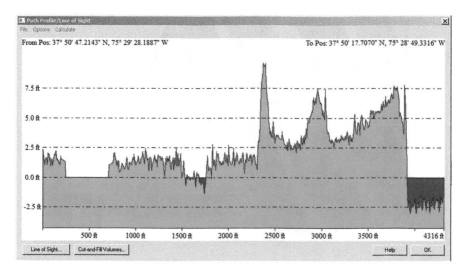

FIGURE 6.7
Terrain profile of the initial DEM.

island. The red diamonds overlaid on the DTM are nesting locations of the endangered piping plover. The inset shows a profile across one of the nesting sites. The beach profile reveals that the piping plover always nests on the leeward side of sand dunes, perhaps to protect their nest from the waves and wind.

Another example of how LiDAR data can be used to study beach habitats is shown in Figure 6.10. The figure shows the habitat locations of three species

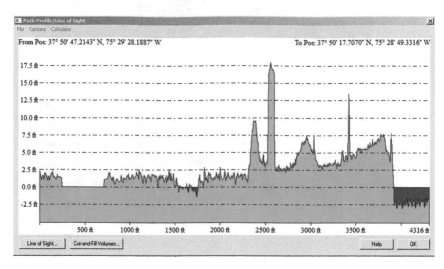

FIGURE 6.8
Terrain profile of the final DTM.

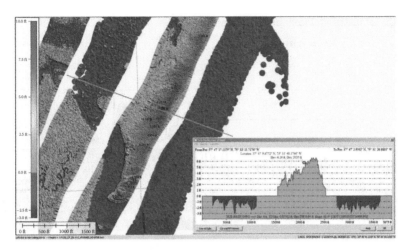

FIGURE 6.9
An example of LiDAR data showing shoreline changes.

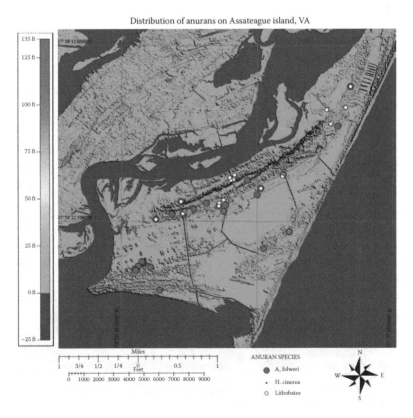

FIGURE 6.10
An example of LiDAR data showing habitat locations of various Anuran species.

of anurans overlaid on a LiDAR-derived topographic surface. Anurans are endangered species that also rely on small elevation changes to build their habitat.

LiDAR maps are frequently used to illustrate small-scale tidal or weather related inundation and coastal flooding events. Figure 6.11a–d demonstrates a simple bathtub model where the different stages of inundation of marshlands and surrounding wetlands around Chincoteague Bay, VA are depicted as hurricane Sandy came ashore. Accurate LiDAR data are important as subtle changes in elevation can affect the horizontal extent of the water surface.

LiDAR elevation accuracy and resolution are important factors when used as input in the Sea Level Affecting Marshes Model (SLAMM) to simulate the dominant processes involved in wetland conversions and shoreline change during long-term SLR. The LiDAR data that are used as input to the model decrease the uncertainty in model prediction.

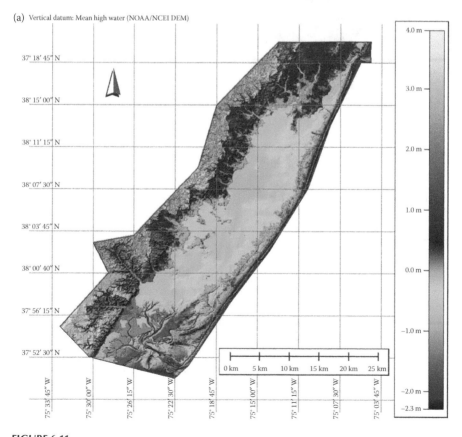

FIGURE 6.11

(a) A DTM of Chincoteague Bay, VA, employing bathymetric LiDAR data. *(Continued)*

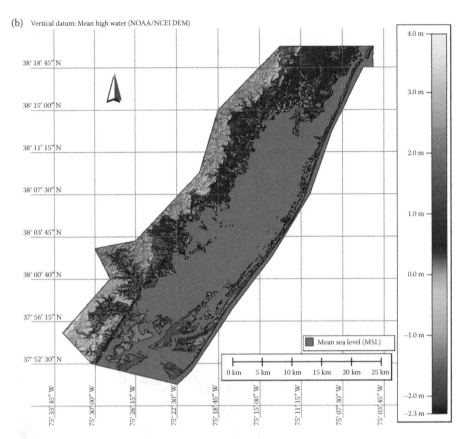

FIGURE 6.11 (Continued)
(b) An inundation model of Mean Sea Level against the NOAA Mean High Water vertical datum, superimposed on the DTM from (a).   *(Continued)*

6.5 Conclusion

Results are discussed from a baseline topographic survey of the NASA WFF and surrounding region. NASA ATM was used to acquire LiDAR data to sample terrain of the aforementioned area. After converting the datasets to sea level elevation, the files were processed using two different approaches and used a variety of software including TerraSolid, geographic information system (GIS), Global Mapper, LP360, and ENVI's LiDAR Toolkits. This research provides a number of DSMs of Wallops and Assawoman Islands in Virginia. The DEMs were furthered classified and processed to produce a number of DTMs. These DEMs and DTMs provide a baseline assessment of Wallops and Assawoman Islands in Virginia and serve to standardize a work flow routine for future DEM and DTM production. Gaps in the

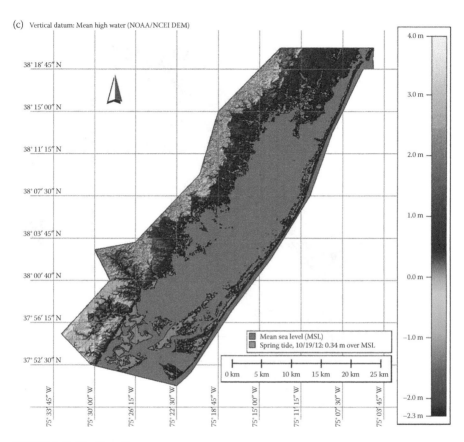

(c) Vertical datum: Mean high water (NOAA/NCEI DEM)

FIGURE 6.11 (Continued)
(c) An inundation model of the Spring Tide from October 2012, superimposed on the DTM from (a).
(*Continued*)

NASA data collection were filled in using data provided by The Nature Conservancy (TNC) DEMs and processed further to provide a complete DTM of the region. The DEM and DTM products were extensively validated using NASA Control points, TNC DEM/DTMs, and field surveys of the area using GPS and Total Station derived ground elevation points. The validation demonstrated that the elevation residuals (DTM minus control point) were less than $+/-14$ cm. This demonstrates that the generated DTMs are better or closely comparable to similar DTMs of the area. Further improvement in the DTMs can be achieved by processing the waveform data supplied by NASA. Further analysis of the data was done using piping plover nesting locations provided by the US Fish and Wildlife Service (USFWS). The results demonstrate that the piping plover uses the leeward side of sand dunes to nest and thus protects its young from offshore physical processes like waves, wind, etc. Although many possible reasons can be gleaned from this result, a complete analysis is needed for confirming the nesting behavior of these birds.

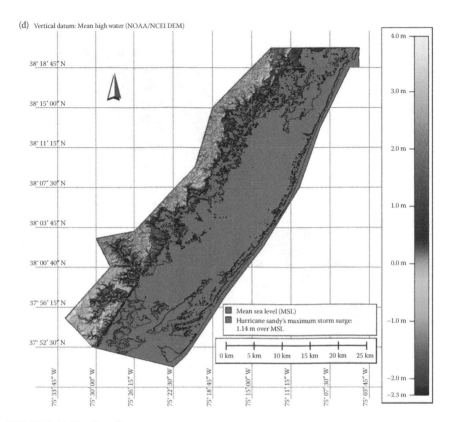

FIGURE 6.11 (Continued)
(d) An inundation model of the maximum storm surge from Hurricane Sandy in October 2012, superimposed on the DTM from (a).

The DTMs also illustrated a number of beach processes that can be observed only from synoptic data collected from remote sensors like the LiDAR. This project not only examined island morphology over time, but also was used in models to study the potential impacts of sea level rise on the coastal ecosystem from climate change and east coast cyclogenesis.

It is the authors' hope that this chapter encourages further applications of LiDAR bare earth models, especially in the marine environment.

Acknowledgments

The authors would like to acknowledge the help and support received from their colleagues at East Stroudsburg University of Pennsylvania and the number of students at Millersville University who helped in the collection, processing, and final rendering of the LiDAR data. This work was supported by funds from the National Aeronautics and Space Administration (NASA) under grant number NNX10AJ10A. Dr. Ajoy Kumar would like to dedicate this work to the memory of Fr. J. Inchackal, S. J.

Bibliography

Axelsson, P. E. 1999. Processing of laser scanner data—Algorithms and applications. *ISPRS Journal of Photogrammetry and Remote Sensing* 54 (2–3): 138–147.

Barbier, N., C. Proisy, C. Vega, and P. Couteron. 2011. Bidirectional texture function of high resolution optical images of tropical forest: An approach using LiDAR hillshade simulations. *Remote Sensing of Environment* 115 (1): 167–179.

Blaunstein, N., S. Aaron, A. Ziberman, and N. Kopeika. 2010. Applied aspects of LiDAR. In *Applied Aspects of Optical Communication and LiDAR*, edited by N. Blaunstein, S. Aaron, A. Ziberman, and N. Kopeika, pp. 123–171. Boca Raton: Auerbach Publications.

Brock, J., C.W. Wright, A. H. Sallenger, W.B. Krabill, and R. Swift. 2002. Basis and methods of NASA's Airborne Topographic Mapper LiDAR Surveys for Coastal Studies. *Journal of Coastal Research* 18 (1): 1–13.

Childs, C. *Interpolating Surfaces in ArcGIS Spatial Analyst.* Retrieved December 20, 2011 from: http://webapps.fundp.ac.be/geotp/SIG/interpolating.pdf.

Craymer, M. 2006. *Making Sense of Evolving Datums: WGS84 and NAD83.* Retrieved June 2, 2011 from: ftp://ftp.geod.nrcan.gc.ca/pub/GSD/craymer/pubs/nad83_hydroscan2006.pdf.

Dietrich, W. E., C. J. Wilson, D. R. Montgomery, and J. McKean. 1993. Analysis of erosion thresholds, channels networks, and landscape morphology using a digital terrain model. *Journal of Geology* 101 (2): 259–278.

Dominguez, R. 2016. *IceBridge DMS L1B Geolocated and Orthorectified Images.* Boulder, Colorado: NASA DAAC at the National Snow and Ice Data Center. http://dx.doi.org/10.5067/OZ6VNOPMPRJ0.

ESRI. 2011. *LiDAR Analysis in ArcGIS 10 for Forestry Applications.* Retrieved June 8, 2011 from ESRI: www.esri.com/library/whitepapers/pdfs/LiDAR-analysis-forestry-10.pdf.

Kirwan, M. L., G. R. Guntenspergen, and J. T. Morris. 2009. Latitudinal trends in *Spartina alterniflors* productivity and the response of coastal marshes to global change. *Global Change Biology* 15 (8): 1982–1989.

Kraak, M. J., F. J. Ormeling, and F. Ormeling. 1996. In *Cartography: Visualization of Spatial Data*, edited by M. J. Kraak, F. J. Ormeling, and F. Ormeling, p. 174. University of Michigan: Longman.

Krabill, W. B., C. W. Wright, R. N. Swift, E. B. Frederick, S. S. Mankade, J. K. Yungel, C. F. Martin, J. G. Sonntag, and M. Duffy. 2000. Airborne laser II. Assateague National Seashore Beach. *Photogrammetric Engineering & Remote Sensing* 66 (1): 65–71.

Krabill, W. B. 2016. *IceBridge ATM L1B Elevation and Return Strength.* Boulder, Colorado: NASA DAAC at the National Snow and Ice Data Center. http://dx.doi.org/10.5067/19SIM5TXKPGT.

Long, T. M., J. Angelo, and J. F. Weishampel. 2011. LiDAR-derived measures of hurri-
      cane and restoration-generated beach morphodynamics in relation to sea turtle
      nesting behavior. *International Journal of Remote Sensing* 32 (1): 231–241.

Mason, D. C., M. Marani et al. 2005. LiDAR mapping of tidal marshes for ecogeomor-
      phological modeling in the TIDE project. *Halifax, Nova Scotia: Eighth International
      Conference on Remote Sensing for Marine and Coastal Environments.*

McCormick, M. P. 2005. Airborne and spaceborne LiDAR. In *LiDAR: Range-Resolved
      Optical Remote Sensing of the Atmosphere*, edited by C. Weitkamp, pp. 355–358.
      New York: Springer Science + Business Media.

Meredith, A., J. Brock et al. 1998. *An Assessment of NASA's Airborne Topographic
      Mapper Instrument for Beach Topographic Mapping at Duck, North Carolina.* Coastal
      Services Center Technical Report CSC/9-98/001.

Morris, J., D. Porter et al. 2005. Integrating LiDAR elevation data, multi-spectral imag-
      ery and neural network modelling for marsh characterization. *International
      Journal of Remote Sensing* 26 (14): 5221–5234.

National Oceanic and Atmospheric Administration (NOAA) Coastal Services Center.
      2012. *Lidar 101: An Introduction to LiDAR Technology, Data, and Applications.*
      Revised. Charleston, SC: NOAA Coastal Services Center.

NOAA Office of Coastal Management. 2017. https://coast.noaa.gov/digitalcoast/.

Nol, E., A. J. Baker, and D. M. Cadman. 1984. Clutch initiation dates, clutch size, and
      egg size of the American Oystercatcher in Virginia. *The Auk* 101 (4): 855–867.

Pfeifer, N. 2008. *Principle and Exploitation of Full Waveform ALS.* Retrieved June 13,
      2011.

Piironon, P. 1994. *A High Spectral Resolution LiDAR Based on an Iodine Absorption
      Filter.* Madison: Madison Space Science and Engineering Center, University of
      Wisconsin.

Roman, C. T., J. A. Peck, J. R. Allen, J. W. King, and P. G. Appleby. 1997. Accretion of
      a New England (U.S.A.) salt marsh in response to inlet migration, storms, and
      sea-level rise. *Estuarine, Coastal and Shelf Science* 45 (6): 717–727.

Sallenger Jr, A. H. et al. 2003. Evaluation of airborne topographic lidar for quantify-
      ing beach changes. *Journal of Coastal Research* 125–133.

Shan, J. and C. Toth. 2009. *Topographic Laser Ranging and Scanning: Principles and
      Processing.* Boca Raton, FL: CRC Press.

Stefan, E. 2010. Doppler wind LiDAR. In *Surface-Based Remote Sensing of the Atmospheric
      Boundary Layer*, edited by E. Stefan, pp. 55–56. Berlin: Springer Netherland.

Stoyanov, D. V., L. L. Gurdev, and T. N. Dreischuh. 1996. Reduction of the pulse
      spike-cut error in Fourier-deconvolved LiDAR profiles. *Applied Optics* 35 (24):
      4798–4802.

Taylor, G., J. Li, D. Kidner, C. Brundson, and M. Ware. 2007. Modelling and predic-
      tion of CPS availability with digital photogrammetry and LiDAR. *International
      Journal of Geographic Information Science* 21 (1): 1–20.

Titus, J. G. 1984. Planning for sea level rise before and after a coastal disaster.
      Greenhouse Effect and Sea Level Rise: A Challenge for This Generation. New York:
      Van Nostrand Reinhold Company.

Titus, J. G. and C. Richman. 2001. Maps of land vulnerable to sea level rise: Modeled
      elevations along the US Atlantic and Gulf coasts. *Climate Research*, 18: 205–228.

Van Den Eeckhaut, D., J. Poesen et al. 2005. The effectiveness of hillshade maps and
      expert knowledge in mapping old deep-seated landslides. *Geomorphology* 67
      (3–4): 351–363.

Vierling, K. T., L. A. Vierling, W. A. Gould, S. Martinuzzi, and R. M. Clawges. 2008. LiDAR: Shedding new light on habitat characteristics and modeling. *Frontiers in Ecology and the Environment* 6 (2): 90–98.

Watershed Sciences. 2010. *LiDAR Remote Sensing Data Collection: Wenas Valley, WA.* Retrieved May 21, 2011 from University of Washington: http://pugetsound-LiDAR.ess.washington.edu/LiDARdata/proj_reports/Wenas_Valley_LiDAR_report.pdf.

Webster, T., D. Forbes, S. Dickie, and R. Shreenan. 2004. Using topographic LiDAR to map flood risk from storm-surge events for Charlottetown, Prince Edward Island, Canada. *Canadian Journal of Remote Sensing* 30 (1): 64–76.

Webster, T., S. Dickie et al. 2003. *Mapping Storm Surge Flood Risk Using a LiDAR-Derived DEM.* Retrieved July 28, 2011, from Elevation: The XYZ's of Digital Elevation Technologies: http://agrg.cogs.nscc.ca/drupal/system/files/documents/project/CCAF%20NB/Webster_et_al_2003_Geospatial.pdf.

White, S. A. and Y. Wang. 2003. Utilizing DEMs derived from LiDAR data to analyze morphologic change in the North Carolina coastline. *Remote Sensing of the Environment* 85 (1): 39–47.

Yohe, G., J. Nuemann, P. Marshall, and H. Ameden. 1996. The economic cost of greenhouse-induced sea-level rise for developed property in the United States. *Climate Change* 32 (4): 387–410.

Zhang, K., S.-C. Chen, D. Whitman, M.-L. Shyu, J. Yan, and C. Zhang. 2003. A progressive morphological filter for removing non-ground measurements from airborne LiDAR data. *Geoscience and Remote Sensing* 41 (4): 872–882.

Zhang, K., D. Whitman, S. Leatherman, and W. Robertson. 2005. Quantification of beach changes caused by Hurricanes Floyd along Florida's Atlantic Coast using airborne laser surveys. *Journal of Coastal Research* 21 (1): 123–134.

7

Spatiotemporal Point Pattern Analysis Using Ripley's K Function

Alexander Hohl, Minrui Zheng, Wenwu Tang,
Eric Delmelle, and Irene Casas

CONTENTS

7.1 Introduction

Many geospatial phenomena are involved with movement across space, for instance, movement of humans and animals, the dispersal of plants, and the diffusion of infectious disease (Gould 1969; Demšar et al. 2015). Our ability to collect fine-resolution spatiotemporal data with high accuracy has substantially improved, given the rapid development of geospatial technologies: location-aware smartphones track their owner's daily movements, sensor networks record biophysical variables in real time, and volunteered geographic information is easily accessible online (Kwan and Neutens 2014). While the study of geospatial movement or diffusion phenomena has received increased attention, the analysis of the spatiotemporal data associated with these phenomena remains challenging (Diggle 2013; Goodchild 2013; An et al. 2015). In this study, we aim to investigate the use of Ripley's K function for the analysis of spatiotemporal point patterns to gain insight into this challenge.

The Ripley's K function (Ripley 1976) is a quantitative approach that falls within the domain of spatial and spatiotemporal point pattern analysis. Spatial point pattern analysis is concerned with quantifying the distribution of point events in 2D geographic space (Illian et al. 2008). It has widespread applications, such as in plant ecology (Wiegand and Moloney 2004; Perry, Miller, and Enright, 2006), epidemiology (Gatrell et al. 1996), and criminology (Anselin et al. 2000). Ripley's K function characterizes a given set of points and distinguishes between random, clustered, and regular patterns. However, the use of point pattern analysis for evaluating spatiotemporally explicit phenomena lags behind in the availability of spatiotemporal datasets. It is important to note that spatiotemporal does not equal 3D, due to the orthogonal relationship between space and time (Nakaya and Yano 2010) and due to the peculiarity of the temporal dimension that clearly distinguishes it from the 2D spatial dimensions (Aigner et al. 2007). This further complicates the analysis of spatiotemporal point patterns.

The objective of this work is to investigate the capability of Ripley's K function-based point pattern analysis for the study of dynamic geospatial phenomena. Specifically, we focus on the combined use of global and local forms of Ripley's K function and present a case study of dengue fever in the city of Cali, Colombia to illustrate the benefits of this methodology.

The remainder of the chapter is organized as follows: Section 7.2 provides the background about global Ripley's K function, followed by Section 7.3, which discusses the temporal extension of the K function and its local variant, as well as specific details about our own implementation of the local Ripley's K function for spatiotemporal point pattern analysis. In Section 7.4, we present a case study where we apply Ripley's K function, followed by results (Section 7.5) and conclusions (Section 7.6).

7.2 Background: Global Ripley's K Function for Spatial Point Pattern Analysis

Different approaches exist to evaluate the level of spatial clustering among point events (Bailey and Gatrell 1995; Delmelle 2009). For instance, the quadrant analysis essentially counts the number of events within each cell (quadrant) of a grid imposed on a study area. The results are compared with the expected frequency of occurrence if the mechanism generating those events was a homogeneous Poisson process (corresponding to point patterns that exhibit complete spatial randomness—i.e., CSR). Despite its ease of implementation, quadrant analysis is an area-based approach that aggregates original point events into quadrant counts, which makes this approach sensitive to the design of quadrants (e.g., shape and size). The nearest neighbor statistic (Diggle, Besag, and Gleaves 1976) alleviates this problem by testing whether point events are closer together (or farther apart) than expected under CSR based on nearest-neighbor distance. This distance-based approach suffers from some limitations, most notably that clustering is typically detected at a relatively small scale, and that distance among events is the only parameter governing the statistic. While the nearest-neighbor approach uses distances only to the closest events and hence only considers the smallest scales of patterns, Ripley's K Function provides a superior alternative in that it evaluates point patterns at different scales (Ripley 1976). By comparing the spatial pattern of the observed data points to simulated data, the K function can indicate for each of the scales evaluated whether the observed point pattern follows a random, clustered, or regular configuration.

Ripley's K function is a statistical approach computed on a set of point events distributed in n-dimensional space, and estimates the second-order property (variance) exhibited by the data. It takes into account (1) the number and (2) distance between the point events, and allows for quantifying how much the observed pattern deviates from randomness at multiple spatial scales (Bailey and Gatrell 1995; Dixon 2013). The theoretical K function, given a set of point-events S, is calculated by dividing E, the number of events that are expected to fall within distance d, by the intensity λ of S (first-order property) as in:

$$K(d) = E(d)/\lambda \qquad (7.1)$$

Equation 7.1 is computed by centering a circle of radius d on each sampling point and counting the number of neighboring events that fall inside it. In this case, the number and locations of the sampling points coincide with the event locations. Dividing the total number of events n by the area of the circle πd^2 results in estimated intensity λ. Ripley's K function is the cumulative distribution of observed point events S with increasing distance. It is expected that $K(d) = \pi d^2$ if the point distribution conforms to CSR, $K(d) > \pi d^2$ if the points cluster within distance d, and $K(d) < \pi d^2$ if the data exhibit a regular pattern. The K function is a second-order analysis of point patterns usually in a two-dimensional space (Haase 1995; Dixon 2013). Second-order effects are caused by the spatial dependence in the process. In essence, Ripley's K function approach uses a circular search window (h) around each event (i) and counts how many other events are observed in that window. The window then moves to the next event until all events (n events) in the study area are visited. The process is repeated at different spatial scales. Specifically,

$$K(h) = (A/n^2) * \sum_{i}^{n} \sum_{j}^{n} (I_h(d_{ij})/w_{ij}) \qquad (7.2)$$

Equation 7.2 evaluates the structural characteristics of a given set of events, where d_{ij} is the distance between events i and j, and A is the size of the study region. The term w_{ij} is a factor to correct for edge effects. The K function is potentially biased as edge effects arise when circles intersect the boundary of the study region. There are different methods to deal with edge effects of the K function, which have been thoroughly studied (Yamada and Rogerson 2003). $I_h(d_{ij})$ is an indicator function defined by Equation 7.3:

$$I_h(d_{ij}) = 1 \quad \text{if} \quad d_{ij} \leq h, \quad 0 \quad \text{otherwise} \qquad (7.3)$$

The K function increases as distance h becomes larger. To statistically test whether the observed point pattern follows a regular, clustered, or random pattern, the K function is evaluated for a large number (M) of Monte Carlo simulations. For each simulation, a number (n) of events are generated (e.g., randomly) within the study area. If the observed K value is larger than the upper simulation envelope, spatial clustering for that distance is statistically significant. Observed K values smaller than the lower simulation envelope show that point patterns exhibit regularity that is statistically significant for the corresponding distance. K(h) is then evaluated against distance (h) to identify the scales at which the point pattern follows a regular, clustered, or random pattern. For a given value of h, if the K function is above, between, or below the upper and lower envelopes, the point pattern is clustered, random, or regular, respectively.

The K function can be transformed to the L function using Equation 7.4 to obtain constant variance with respect to a benchmark of zero, which facilitates the comparison of L values across all h:

$$L(h) = (K(h)/\pi)^{1/2} - h \qquad (7.4)$$

where $L(h) = 0$ if the pattern conforms to CSR, $L(h) > 0$ if clustered, and $L(h) < 0$ for regular patterns.

Recent methodological advancements of Ripley's K function improve the ability to find the appropriate scale of clustering by computing K for distance increments (Tao, Thill, and Yamada 2015). In addition, the K function was adapted for network-constrained data (Yamada and Thill 2007) which violate the planar space assumption that underlies many spatial point pattern analysis methods. Although the K function is popular because it evaluates levels of clustering at different scales, its computation is time-consuming, especially for large datasets. Recent work has underscored the capability of high performance computing, for instance through the use of Graphics Processing Units (GPU) for its acceleration (Tang, Feng, and Jia 2015).

7.3 Ripley's K Function for Spatiotemporal Point Pattern Analysis

In this study, we investigate the use of Ripley's K function in the analysis of spatiotemporal point patterns. We focus on the combination of global and local forms of Ripley's K function. While the former evaluates the spatiotemporal characteristics of a point pattern at the aggregated level (i.e., the entire dataset), the local form of Ripley's K function quantifies the characteristics of the point pattern, as well as its deviation from what would be expected, locally. In this section, we discuss in detail global and local forms of Ripley's K function for spatiotemporal point pattern analysis.

7.3.1 Global Ripley's K Function for the Analysis of Spatiotemporal Point Pattern

Tests for spatial patterns fail at evaluating the dynamics of the point process. When point events have a temporal attribute, we can investigate whether two events are space and time dependent, suggesting the presence of a space-time link. There are several techniques to evaluate patterns among spatiotemporal point events. The Knox test for space-time interaction evaluates the presence of a space-time cluster at given spatial and temporal distances (Knox 1964). Knox's test method is limited due to its arbitrary definition of closeness and the critical distance does not account for population heterogeneity (Jacquez 1996). The Mantel test (Mantel 1967) incorporates the notion of distance decay in which nearby pairs of events are more important than distant pairs. Jacquez's k-Nearest Neighbor k-NN statistic (Jacquez 1996) addresses the weaknesses of the Knox and Mantel statistics by counting the number of pairs of events that are nearest neighbors in both space and time.

Equivalent to the purely spatial K function (discussed above), the space-time Ripley's K function is a global statistic computed on the entire set of space-time point events. The theoretical space-time K function, given a set of point events S, is calculated by dividing E, the number of events that are expected to fall within spatial distance d and temporal distance t, by the intensity λ of S (first-order property):

$$K(d, t) = E(d, t) / \lambda \qquad (7.5)$$

Equation 7.5 characterizes the pattern of S within the space-time cube framework (Nakaya and Yano 2010), where a cylinder of base πd^2 and height t is centered on each sampling point to compute the number of events falling within. Again, the sampling points are equal to the event locations. Then the total number of events n is divided by the volume of the irregular prism formed by the study area/period, which results in intensity λ. Space-time Ripley's K function is the cumulative distribution of observed point events S with increasing space and time distance. It is expected that $K(d,t) = \pi d^2 t$ if the point distribution conforms to complete spatiotemporal randomness (CSTR), $K(d,t) > \pi d^2 t$ if the points cluster within spatial and temporal distance d and t, and $K(d,t) < \pi d^2 t$ for regular space-time patterns. Using Equation 7.6, the space-time K function is formulated as (Bailey and Gatrell 1995):

$$K(h,t) = ((L*R)/n^2)*\sum_{i}^{n}\sum_{j}^{n}(I_{h,t}(d_{ij},t_{ij})/w_{ij}) \qquad (7.6)$$

where t_{ij} is the time that separates two events i and j. d_{ij} is the distance between events i and j. L denotes the area of the study region and R is the duration of the study period. The product of L and R results in the volume of the irregular prism that is formed by the study area (base) and the study period (height). $I_{h,t}(d_{ij},t_{ij})$ is an indicator function defined in Equation 7.7:

$$I_{h,t}(d_{ij},t_{ij}) = 1 \quad \text{if} \quad d_{ij} \leq h \quad \text{and} \quad t_{ij} \leq t, 0 \quad \text{otherwise} \qquad (7.7)$$

Larger time t and distance h intervals will contribute to an increase in the space-time K function. For the case that no space-time interaction exists, Equation 7.6 becomes the product of the spatial and temporal K functions K(h)*K(t). Testing for space-time dependence is achieved by subtracting K(h)*K(t) from the combined space-time I function K(h,t). Methods for handling edge effects of the space-time K function have been studied by Gabriel (2014).

The space-time Ripley's K function is transformed to the space-time L function by Equation 7.8:

$$L(h,t) = (K(h,t)/\pi t)^{1/2} - h \qquad (7.8)$$

where $L(h,t) = 0$ under CSTR, $L(h,t) > 0$ for clustered patterns, and $L(h,t) < 0$ for regular patterns.

7.3.2 Local Ripley's K Function for the Analysis of Spatiotemporal Point Patterns

While Ripley's K evaluates the spatial pattern at the global level (i.e., indicating whether clustering is present in the entire study area or not), the same measure can be considered in its local form to pinpoint where the clustering actually occurs (Getis and Franklin 1987):

$$K_i(h) = (A/n) * \sum_{j}^{n} (I_h(d_{ij})/w_{ij}) \qquad (7.9)$$

Here, the local K function is evaluated at each sampling point i, which either is part of (1) a regularly spaced grid drawn over the study area, (2) the events themselves, or (3) a number of random points. The indicator function I is equivalent to Equation 3. Several meaningful extensions to local Ripley's K function have been suggested, such as the local K function for network-constrained space to study transportation-related cases (Okabe and Yamada 2001; Yamada and Thill 2007), as well for characterizing patterns in flow data, thereby upgrading the classic hot spot detection paradigm to the stage of "hot flow" detection (Tao and Thill 2016).

Equivalent to the purely spatial case, the local K function for spatiotemporal point patterns identifies the location and time of clusters within the study area/period. The local space-time K function is evaluated at any sampling point i:

$$K_i(h,t) = ((L*R)/n^2) * \sum_{j}^{n} (I_{h,t}(d_{ij},t_{ij})/w_{ij}) \qquad (7.10)$$

Using Equation 7.10 for each sampling point i, we can estimate the local level of space-time clustering and its statistical significance using Monte Carlo simulations. Further, we can identify the scale at which space-time clustering is the greatest. This information can be very valuable when conducting spatial analysis over a non-homogeneous population of events. Despite these attractive outcomes, the local version of the space-time K function is computationally very demanding, and the execution time depends on (1) the number of data points, (2) the number of sampling points, and (3) the number of spatial and temporal bandwidths, for which K is computed.

7.4 Case Study

To gain insights into the mechanisms of spatiotemporal point pattern analysis, we now illustrate our implementation of the global and local Ripley's K function spatiotemporal explicit set of dengue fever cases in Colombia for the years 2010–2011.

7.4.1 Study Area and Data

The city of Cali is located in the southwest of Colombia. It is the third largest metropolitan area in the country with a total population of around 2.3 million and a population density of $4140/km^2$ in 2013 (Cali 2014). Cali experiences two rainy seasons: the first from April to July and the second from September to December. Located at approximately 1000 m above sea level, it has an average temperature of 26°C and an average precipitation of 1000 mm over most of the metropolitan area (Cali 2014). The city is administratively divided into 22 communes covering 120.9 km^2, and composed of 340 neighborhoods (see Figure 7.1). A commune is a grouping of neighborhoods based on homogeneous demographic and socioeconomic characteristics. Neighborhoods are classified using a stratification system composed of six classes, one being the lowest and six the highest. The strata are developed by evaluating the type of housing, urban environment, and context. The city, as in most colonial cities in Latin America, grew from its central core, following the city spine, and toward the periphery. Peripheral neighborhoods are typically characterized

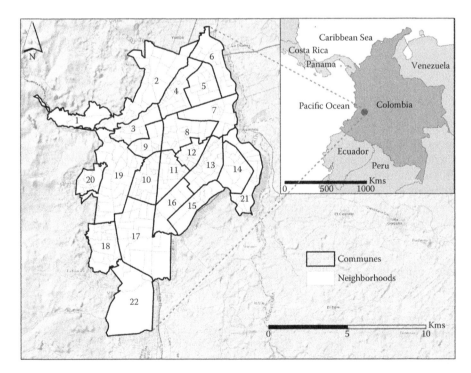

FIGURE 7.1
The city of Cali, Colombia. (Esri, HERE, DeLorme, Intermap, increment P Corp., GEBCO, USGS, FAO, NPS, NRCAN, GeoBase, IGN, Kadaster NL, Ordnance Survey, Esri Japan, METI, Esri China (Hong Kong), swisstopo, MapmyIndia, © OpenStreetMap contributors, and the GIS User Community.)

by high density and low income since they have been the result of squatter settlements and poor urban planning (Restrepo 2011).

We use a dataset of dengue fever cases within the city of Cali in this study. The data are extracted from the "Sistema de Vigilancia en Salud Pública (SIVIGILA)" (English: Public Health Surveillance System) for the city of Cali for the years 2010 and 2011. The SIVIGILA system has as a main responsibility to observe and analyze health events with the objective of planning, follow-up, and evaluation of public health practices (Colombia 2017). Reported cases of dengue fever are entered into the system daily. Each case includes personal information about the patient such as their home address and when they were diagnosed. A total of 11,056 cases were geocoded to the closest intersection to guarantee a level of privacy for both years. There were 9606 cases in 2010 and 1562 in 2011. The difference in the number of cases is explained by the fact that 2010 was identified as an epidemic year (Varela et al., 2010).

7.4.2 Analysis

7.4.2.1 Global Space-Time K Function

Since the epidemiological interest is to find clusters of disease occurrence, we evaluated the magnitude of space-time clustering within the dengue fever dataset (n = 11,056) by computing the global space-time Ripley's K and corresponding L functions (see Section 3.1). We used spatial bandwidths from 50 m to 1000 m in 50 m increments and temporal bandwidths from 0 to 14 days in 1-day increments. Using Equation 7.11, we assessed statistical significance of the observed K function by comparison with 100 population-adjusted random simulations and finding the spatial and temporal scales at which the difference between the observed function and the upper simulation envelope (noted as $L_{diff\_upper}(h,t)$) was maximal (also see Hohl et al. 2016):

$$L_{diff\_upper}(h,t) = L_{obs}(h,t) - L_{upper\_envelope}(h,t) \qquad (7.11)$$

where $L_{obs}(h,t)$ is the observed L value and $L_{upper\_envelope}(h,t)$ represents the L value of the upper simulation envelope at spatial bandwidth h and temporal bandwidth t.

7.4.2.2 Local Space-Time K Function

Once we determined the presence of clusters in the dengue fever dataset, we illustrated the locations and times at which the clusters may occur by computing the local space-time Ripley's K function. We imposed a regularly spaced grid of sampling points on our study area/period using a space-time resolution of 250 m and 7 days. This results in a total of 202,755 sampling points at which the local space-time K functions were evaluated (although more accurate results can be obtained at a finer scale, e.g., 100 m and 1 day,

estimating local space-time clusters every 250 m and 7 days is computationally more accessible). Equivalent to our estimation of the global space-time
K function (see Section 4.2.2), we used spatial bandwidths of 50 m to 1000 m
in increments of 50 m and temporal bandwidths of 0–14 days in 1-day increments at each sampling point. Again, using Equation 11, we assessed statistical significance of the observed local K function by comparison with the
upper simulation envelope of 100 population-adjusted Monte Carlo simulations. To illustrate the effects of scale on space-time point pattern, we show
significantly clustered sampling locations at two different scales by drawing
a point cloud within the space-time cube (Delmelle et al. 2014): (1) 500 m and
7 days, (2) 750 m and 10 days.

7.4.2.3 Global Space-Time K Function of Local Settings

For illustration purposes, we assess the magnitude and statistical significance
of space-time clustering at various scales by selecting three distinct locations
from the space-time grid of sampling points (see section 4.2.2). Each of the
three locations is representative of a particular space-time pattern. We chose
Location 1 in the center of the dengue fever cluster in the south-western part
of the city during the first half of 2010. Location 2 is the same as Location 1,
but has a much later time stamp, during which the infectious outbreak is
in its declining stage. It can be seen that Locations 1–2 are sites where the
virus is present throughout the endemic period. It is a constant focal point
of infection for more than 150 days. This information is valuable to health
authorities in order to target the location to stop the spread of the disease.
This area corresponds to a military base where the municipality spraying
cycles are not as regular as in other areas in the city. Location 3 lies in the
eastern part of the city, which never exhibits a clustered pattern during the
entire study period. Their space-time coordinates (x, y, t), using the Bogota
Transverse Mercator coordinate system and Julian date [0–730], are: Location
1: (1,058,498.1, 864,811.9, 35); Location 2: (1,058,498.1, 864,811.9, 210); Location 3:
(1,064,998.1, 870,311.9, 35). For each of the three locations, we identified surrounding dengue fever cases within distance 1000 m and 14 days, and computed the global space-time K and L functions for this local setting (the same
way as we compute the global K and L functions for the entire study area/
period as in Section 4.2.1). To distinguish between clustered, random, and
regular space-time patterns, we compared the observed L functions with an
upper simulation envelope from the population adjusted simulations using
Equation 11, as well as a lower simulation envelope using Equation 7.12:

$$L_{\text{diff\_lower}}(h,t) = L_{\text{obs}}(h,t) - L_{\text{lower\_envelope}}(h,t) \qquad (7.12)$$

where $L_{\text{obs}}(h,t)$ is the observed L value and $L_{\text{lower\_envelope}}(h,t)$ represents the L
value of the lower simulation envelope at spatial bandwidth h and temporal
bandwidth t.

7.4.2.4 Implementation

All programs were written using Python and R (package: stpp) and we used a high performance computing cluster with 32 nodes connected through an infiniband network switch (Pfister, 2001) to accelerate the spatiotemporal point pattern analysis using Ripley's K function. Each computing node of the high performance computing cluster has 12 CPUs and 12 GBs of memory, in total 384 CPUs (Intel Xeon processor with a 2.67 GHz clock speed). Similar to Delmelle et al. (2014), we used Voxler, an interactive 3D modeling environment (Golden Software, Colorado), for the visualizations of the local space-time K functions.

7.5 Results

7.5.1 Global Space-Time K Function

Figure 7.2 shows the difference between L values of the observed data and the upper simulation envelope of 100 population-adjusted Monte Carlo

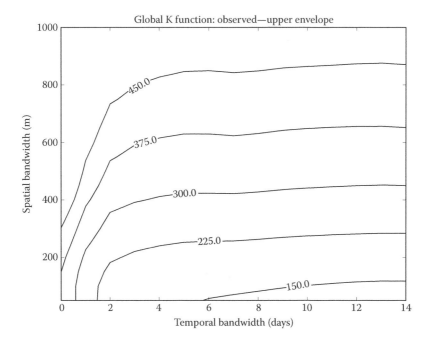

FIGURE 7.2
Global space-time K function. Absolute difference between observed L values and upper simulation envelope.

simulations. Note that we did not account for edge effects in our case study, therefore setting parameter w_{ij} to 1 in all calculations (same for the K function analysis of local settings). Observed L values greater than simulated ones (positive difference) indicate clustering at the corresponding scale: the greater the difference, the stronger the clustering is. Figure 7.2 shows positive values across all bandwidths, with the overall trend of stronger clustering for larger spatial bandwidths. However, the magnitude of the difference in clustering decreases with increasing temporal bandwidth, suggesting that dengue fever cases tend to occur shortly after one another, but do not exhibit strong temporal clustering beyond a week. Thus, the change in clustering intensity is mainly driven by the spatial scale, meaning that, as opposed to changing the temporal bandwidth, the difference becomes larger when increasing the spatial bandwidth.

7.5.2 Local Space-Time K Function

We estimate the local space-time K function at 202,764 regularly spaced grid points (250 m, 7 days intervals). For each grid point, we report the absolute difference in L values of the observed data and the upper simulation envelope of 100 population-adjusted Monte Carlo simulations at different spatial and temporal bandwidths.

Figure 7.3 illustrates a map that visualizes the difference in L-values for each grid point (or voxel, volumetric pixel), using a spatial and temporal bandwidth of 500 m and 7 days (from two different perspectives: southeast and northwest). Negative values (where observed counts are less than expected) are not shown on the map. Colored dots denote regions where the number of observed cases is greater than what is expected; the magnitude of this difference is illustrated with tones of red (darker red dots are on the end

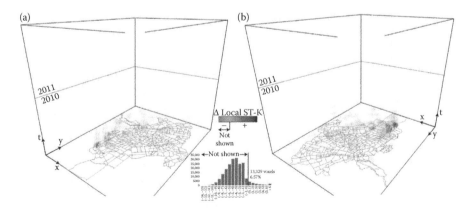

FIGURE 7.3
Local space-time K function using spatial and temporal bandwidths of 500 m and 7 days, respectively. Only positive differences between observed data and the upper simulation envelope are shown (a) from the Southeast, and (b) from the Northwest.

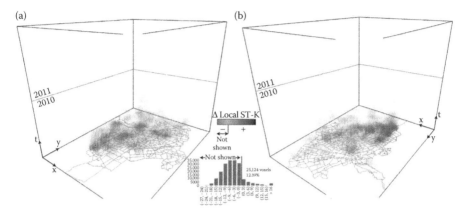

FIGURE 7.4
Local space-time K function using spatial and temporal bandwidths of 750 m and 10 days, respectively. Only positive differences between observed data and the upper simulation envelope are shown (a) from the Southeast, and (b) from the Northwest.

of that spectrum). We note the presence of strong clusters at the beginning of the year 2010, coinciding with an increase in cases during the first few months of the year (see Hohl et al. 2016). Figure 7.4 is similar to Figure 7.3, but uses larger spatial and temporal bandwidths (750 m and 10 days). Using the same legend as in Figure 7.3, we observe a much greater number of grid points where the difference between the observed and expected L-values is large (n = 25,124 voxels or 12.39%, compared to 13,329 or 6.57% in the former scenario with spatial and temporal bandwidths of 500 m and 7 days).

Figure 7.5 shows the interpolated variation in the space-time K function, with bandwidths of 500 m and 7 days. Essentially, this map uses data from

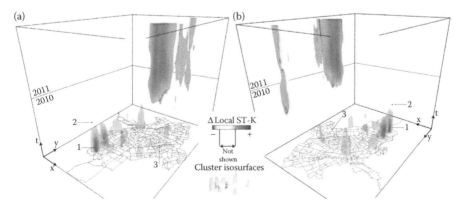

FIGURE 7.5
Local space-time K function using spatial and temporal bandwidths of 500 m and 7 days, respectively. Absolute difference between observed data and the upper simulation envelope. Locations 1, 2, and 3 for global space-time K functions of local settings are annotated in the map (a) from the Southeast, and (b) from the Northwest.

Figure 7.3 as an input, but shows a smooth continuous volume. We use a combination of visualization techniques (volume rendering, transparency; see Delmelle et al. (2014)) to render the strength of the clustering, while iso-surfaces reinforce the extent of such clusters. For this map, of more interest are regions of strong positive clustering (represented in red). Although we observe strong, positive clusters at the beginning of 2010, we note that at the end of 2011, some regions are showing negative values, suggesting a tendency toward regularity.

7.5.3 Global Space-Time K Functions of Local Settings

Figure 7.6 depicts the absolute difference between observed Ripley's K and the upper simulation envelope for Location 1 within 1000 m and 14 days of the space-time bandwidth. Location 1 lies within a space-time cluster of dengue fever cases. Local Ripley's K values suggest clustered patterns of dengue fever cases within 1000 m and 14 days with respect to Location 1 since all the values of the absolute difference with respect to the upper simulation envelope are positive. As the spatial bandwidth increases, the clustering response

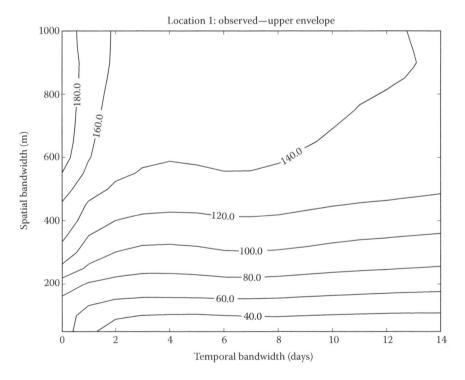

FIGURE 7.6
Global space-time K function of a local setting (Location 1 in Figure 7.5). Absolute difference between observed data and the upper simulation envelope.

becomes stronger, but weakens with decreasing temporal bandwidth values. Strongest clustering response concentrates within a space-time range of 500–1000 m and 1 day.

The difference between observed data and the upper simulation envelope for Location 2 is shown in Figure 7.7. For this location, there is no clustering pattern that we can observe. However, when the temporal bandwidth is from 0 to 1 day or 6 to 14 days within 150 m of the spatial bandwidth, the dengue fever data around that location exhibit a weaker clustering pattern—only a small-scale cluster is observed. To have a better understanding of the dengue fever pattern at Location 2, we compare the difference between observed data and the lower simulation envelope (see Figure 7.8). When the difference between the observed and simulated data is higher than the lower envelope, a completely spatiotemporally random (CSTR) pattern is suggested. Otherwise, space-time regularity is observed when the difference with respect to the lower envelope is negative. As we see in Figure 7.8, positive values of the difference with the lower envelope are observed across all space-time bandwidths. Therefore, we cannot reject the null hypothesis that the spatiotemporal pattern of dengue fever incidents is completely random.

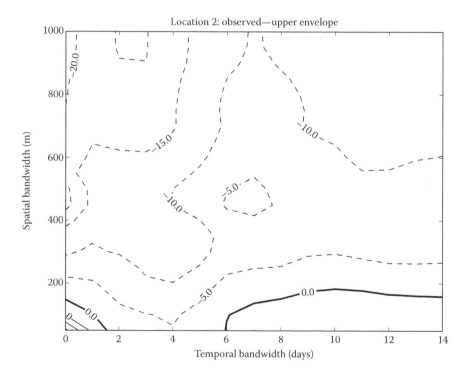

FIGURE 7.7
Global ST K function of a local setting (Location 2 in Figure 7.5). Absolute difference between observed data and the upper simulation envelope. Dashed lines denote negative differences.

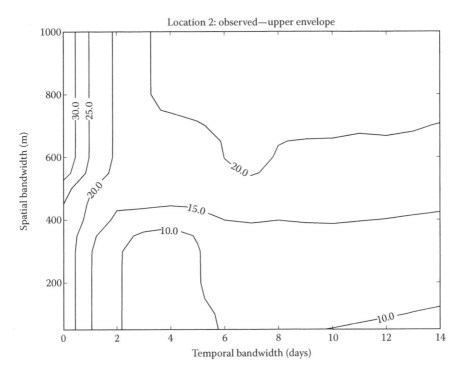

FIGURE 7.8
Global space-time K function of a local setting (Location 2 in Figure 7.5). Absolute difference between observed data and the lower simulation envelope.

For Location 3, all the differences in L-values between observed and upper simulation envelope are negative (see Figure 7.9). When we plot the difference between observed values and the lower simulation envelope (see Figure 7.10), we observe a random pattern throughout all scales, especially for spatial bandwidths from 150 to 750 m and temporal bandwidths between 0 and 1 day. For spatial bandwidths from 250 to 650 m and temporal bandwidths longer than 1 day, the dengue fever data exhibit spatiotemporal regularity with respect to Location 3.

7.6 Conclusions

In this study, we investigated the use of Ripley's K function for the analysis of spatiotemporal point patterns. Using a combination of global and local Ripley's K functions allowed us to discover the space-time characteristics of dengue fever in Cali, Colombia for the years 2010 and 2011.

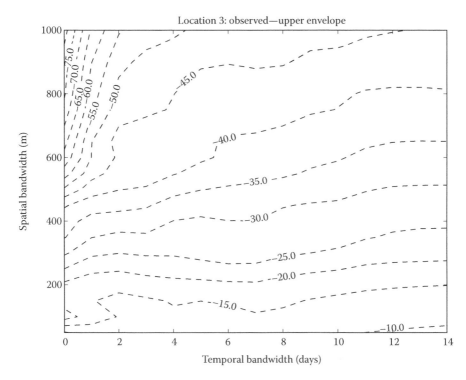

FIGURE 7.9
Global space-time K function of a local setting (Location 3 in Figure 7.5). Absolute difference between observed data and the upper simulation envelope. Dashed lines denote negative differences.

In the case of dengue fever and other vector-borne diseases, being able to identify the space-time location of potential clusters of infection can make a difference in controlling and stopping the spread of the virus. It will help health authorities to better design and plan control strategies in a timely way to stop an epidemic from happening. It will also provide insight into understanding the timeline of the infectious process.

The 3D visualization approach is able to map the shape of each cluster, while giving a clear understanding of the presence of clusters of dengue fever over space and time. Our future work will focus on a number of threads. First, we will perform edge correction to improve global and local forms of space-time Ripley's K function. Second, once a fine spatial-temporal resolution is used, the 3D visualization approach will map more accurate shapes and forms of each cluster. Third, more years of dengue fever data will be added in our study to better understand and explain the space-time complexity of the infectious process.

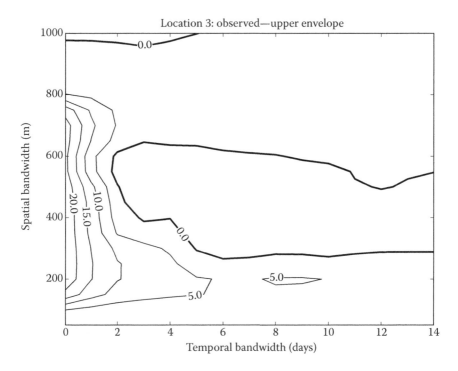

FIGURE 7.10
Global space-time K function of a local setting (Location 3 in Figure 7.5). Absolute difference between observed data and the lower simulation envelope.

Acknowledgment

The authors would like to thank the Public Health Secretariat of the city of Cali for the dengue fever surveillance data. Support of computing resources from University Research Computing (URC) at the University of North Carolina at Charlotte, U.S. NSF XSEDE Supercomputing Resource Allocation (SES170007) is acknowledged.

References

Aigner, W., S. Miksch, W. Müller, H. Schumann, and C. Tominski. 2007. Visualizing time-oriented data—A systematic view. *Computers & Graphics* 31 (3): 401–409.

An, L., M.-H. Tsou, S. E. Crook, Y. Chun, B. Spitzberg, J. M. Gawron, and D. K. Gupta. 2015. Space–time analysis: Concepts, quantitative methods, and future directions. *Annals of the Association of American Geographers* 105 (5): 891–914.

Anselin, L., J. Cohen, D. Cook, W. Gorr, and G. Tita. 2000. Spatial analyses of crime. *Criminal Justice* 4 (2): 213–262.

Bailey, T. and Q. Gatrell. 1995. *Interactive Spatial Data Analysis*. Edinburgh Gate, England: Pearson Education Limited.

Cali, A. d. S. d. ed. 2014. *Cali en cifras*. Cali, Colombia: Alcaldía de Santiago de Cali.

Chen, B. Y., H. Yuan, Q. Li, S.-L. Shaw, W. H. Lam, and X. Chen. 2016. Spatiotemporal data model for network time geographic analysis in the era of big data. *International Journal of Geographical Information Science* 30 (6): 1041–1071.

Colombia, M. d. S. 2017. Sistema de vigilancia en salud pública. Ministerio de Salud Colombia 2017 [cited February 2017]. Available from https://www.minsalud. gov.co/salud/Paginas/SIVIGILA.aspx.

Delmelle, E. 2009. Point pattern analysis. *International Encyclopedia of Human Geography* 8: 204–211.

Delmelle, E., C. Dony, I. Casas, M. Jia, and W. Tang. 2014. Visualizing the impact of space-time uncertainties on dengue fever patterns. *International Journal of Geographical Information Science* 28 (5): 1107–1127.

Demšar, U., K. Buchin, F. Cagnacci, K. Safi, B. Speckmann, N. Van de Weghe, D. Weiskopf, and R. Weibel. 2015. Analysis and visualisation of movement: An interdisciplinary review. *Movement Ecology* 3 (1): 5.

Diggle, P. J. 2013. *Statistical Analysis of Spatial and Spatio-Temporal Point Patterns*. Boca Raton: CRC Press.

Diggle, P. J., J. Besag, and J. T. Gleaves. 1976. Statistical analysis of spatial point patterns by means of distance methods. *Biometrics* 32: 659–667.

Dixon, P. M. 2013. *Ripley's K function*. Encyclopedia of Environmetrics.

Gabriel, E. 2014. Estimating second-order characteristics of inhomogeneous spatiotemporal point processes. *Methodology and Computing in Applied Probability* 16 (2): 411–431.

Gatrell, A., T. Bailey, P. Diggle, and B. Rowlingson. 1996. Spatial point pattern analysis and its application in geographical epidemiology. *Transactions of the Institute of British Geographers* 21 (1): 256–274.

Getis, A. 1984. Interaction modeling using second-order analysis. *Environment and Planning A* 16 (2): 173–183.

Getis, A. and J. Franklin. 1987. Second-order neighborhood analysis of mapped point patterns. *Ecology* 68 (3): 473–477.

Goodchild, M. F. 2013. Prospects for a space–time GIS: Space–time integration in geography and GIScience. *Annals of the Association of American Geographers* 103 (5): 1072–1077.

Gould, P. R. 1969. *Spatial Diffusion. Resource Paper 4*. Washington, DC: Association of American Geographers.

Haase, P. 1995. Spatial pattern analysis in ecology based on Ripley's K-function: Introduction and methods of edge correction. *Journal of Vegetation Science* 6 (4): 575–582.

Hohl, A., E. Delmelle, W. Tang, and I. Casas. 2016. Accelerating the discovery of space-time patterns of infectious diseases using parallel computing. *Spatial and Spatio-Temporal Epidemiology* 19: 10–20.

Illian, J., A. Penttinen, H. Stoyan, and D. Stoyan. 2008. *Statistical Analysis and Modelling of Spatial Point Patterns*. New York: John Wiley & Sons.

Jacquez, G. M. 1996. A k nearest neighbour test for space–time interaction. *Statistics in Medicine* 15 (18): 1935–1949.

Kiskowski, M. A., J. F. Hancock, and A. K. Kenworthy. 2009. On the use of Ripley's K-function and its derivatives to analyze domain size. *Biophysical Journal* 97 (4): 1095–1103.

Knox, G. E. 1964. The detection of space-time iterations. *Journal of the Royal Statistical Society* 13: 25–29.

Kwan, M.-P. and T. Neutens. 2014. Space-time research in GIScience. *International Journal of Geographical Information Science* 28 (5): 851–854.

Mantel, N. 1967. The detection of disease clustering and a generalized regression approach. *Cancer Research* 27 (2 Part 1): 209–220.

Mountrakis, G. and K. Gunson. 2009. Multi-scale spatiotemporal analyses of moose–vehicle collisions: A case study in northern Vermont. *International Journal of Geographical Information Science* 23 (11): 1389–1412.

Nakaya, T. and K. Yano. 2010. Visualising crime clusters in a space-time cube: An exploratory data-analysis approach using space-time kernel density estimation and scan statistics. *Transactions in GIS* 14 (3): 223–239.

Okabe, A., B. Boots, and T. Satoh. 2010. A class of local and global K functions and their exact statistical methods. In *Perspectives on Spatial Data Analysis*, edited by L. Anselin, and S. Rey, 101–112. Berlin: Springer.

Okabe, A., and I. Yamada. 2001. The K-Function Method on a Network and Its Computational Implementation. *Geographical Analysis* 33 (3): 271–290.

Perry, G. L., B. P. Miller, and N. J. Enright. 2006. A comparison of methods for the statistical analysis of spatial point patterns in plant ecology. *Plant Ecology* 187 (1): 59–82.

Pfister, G. F. 2001. An introduction to the infiniband architecture. *High Performance Mass Storage and Parallel I/O* 42: 617–632.

Restrepo, L. D. E. 2011. El plan piloto de cali 1950. *Bitácora Urbano Territorial* 1 (10): 222–233.

Ripley, B. D. 1976. The second-order analysis of stationary point processes. *Journal of Applied Probability* 13 (2): 255–266.

Tang, W., W. Feng, and M. Jia. 2015. Massively parallel spatial point pattern analysis: Ripley's K function accelerated using graphics processing units. *International Journal of Geographical Information Science* 29 (3): 412–439.

Tao, R. and J. C. Thill. 2016. Spatial Cluster Detection in Spatial Flow Data. *Geographical Analysis* 48 (4): 355–372.

Tao, R., J.-C. Thill, and I. Yamada. 2015. Detecting clustering scales with the incremental K-function: comparison tests on actual and simulated geospatial datasets. In *Information Fusion and Geographic Information Systems (IF&GIS'2015)*, edited by V. Popovich, C. Claramunt, M. Schrenk, K. Korolenko, and J. Gensel, 93–107. Heidelberg: Springer.

Varela, A., E. G. Aristizabal, and J. H. Rojas. 2010. *Analisis epidemiologico de dengue en Cali*. Cali: Secretaria de Salud Publica Municipal.

Wiegand, T. and K. A. Moloney. 2004. Rings, circles, and null-models for point pattern analysis in ecology. *Oikos* 104 (2): 209–229.

Yamada, I. and P. A. Rogerson. 2003. An Empirical Comparison of Edge Effect Correction Methods Applied to K-function Analysis. *Geographical Analysis* 35 (2): 97–109.

Yamada, I. and J. C. Thill. 2007. Local Indicators of Network-Constrained Clusters in Spatial Point Patterns. *Geographical Analysis* 39 (3): 268–292.

Yamada, I. and J.-C. Thill. 2010. Local indicators of network-constrained clusters in spatial patterns represented by a link attribute. *Annals of the Association of American Geographers* 100 (2): 269–285.

8

Geospatial Data Science Approaches for Transport Demand Modeling

Zahra Navidi, Rahul Deb Das, and Stephan Winter

CONTENTS

8.1 Introduction

Modeling transport—the phenomenon of people or goods moving in vehicles in space and time and being constrained in that movement by transport networks—is inherently related to geospatial data (Miller and Shaw 2001, 2015). People, goods, and vehicles are located somewhere at any time, in relation to each other as well as in relation to various mode-specific transport networks; they are coming from some location and heading to another location within some time constraints. The movements of people and goods—collectively defining the transport demand—can be solitary or shared, but are neither independent of each other, nor independent of the vehicles, which are defining the transport supply. In addition to the factors of space and time, economic, social, and individual factors also determine choice and behavior. That is why transport has long been recognized as a complex system and, as such, hard to model.

Nevertheless, transport models are needed for a variety of reasons, from traffic management to environmental or economic hypothesis testing. Furthermore, the availability of data about and from transport provides ample information for travelers to make individual decisions, and thus, become smarter about mobility. Given the dynamic nature and the complexity arising from the behavior of many actors, the established method for transport modeling is agent-based simulation (Torrens 2004). Agent-based simulation relies on transport demand modeling, which relies again in a complex way on geospatial data, including georeferenced sociodemographic data, economic data, and environmental data. This chapter will disentangle this relationship, and review state-of-the-art methods of transport demand modeling for agent-based transport simulations.

The focus of this chapter is on the key role of geospatial data in improving smart mobility. Firstly, the categories of transport-related geospatial data and surveys are introduced and explained. Secondly, demand modeling—from the traditional four-step model to activity-based modeling—and the required datasets are reviewed. Thirdly, the challenges and issues of synthesizing a population and creating a population-wide demand model based on surveys and geospatial data are reviewed. Finally, by explaining and illustrating the key role of demand models in agent-based transport simulation and mobility planning, the importance of geospatial data and their application in the transport domain will be elaborated.

8.2 Technological Evolution of Travel Data Collection

Data related to travel behavior is the most crucial yet limiting aspect of transportation modeling (McNally and Rindt 2008). Transportation modeling is a complex problem that requires a wide range of data types such as spatial, temporal, user sociodemographic characteristics, and qualitative data about users' experience. Stopher (2008) classified the required quantitative data into two categories of supply data (e.g., capacity, design speed, and type of services provided) and demand data (e.g., volume of demand, users' characteristics, and demand density), and emphasized that, depending on the objectives of each study, various qualitative data also may be required for collection.

8.2.1 Manual Travel Surveys

Travel surveys are the main method of data collection in the transportation field. In the current state of the art, there is a considerable human effort involved in travel surveys. In terms of participation, a travel survey can be categorized into two main types: participatory and non-participatory. A participatory survey requires the active participation of the users, either direct (filling the forms themselves) or indirect (answering to someone who is filling the form), whereas a non-participatory survey involves observing the travel behavior through roadside volunteers who count the number of travelers in a given transport mode or monitor the transport resources in a given network.

In this regard, non-participatory surveys are mainly deployed to collect supply data, such as: traffic counting, network inventory, and land-use inventory (Stopher 2008).

On the other hand, the bulk of demand data is collected through participatory surveys, which, owing to the scale of application, requires a huge amount of time and money. There are other complications that hinder the procedure of demand data collection in the transportation field as well. For example, people are concerned about protecting their privacy; they do not like to be followed by cameras or answer surveys that reveal their personal daily plans and destinations.

Household travel surveys (HTS) are the most common survey to collect travel and activity data from a representative sample of the population in a given region, typically over a period of 24 hours of a workday to an even longer duration. The surveys can also take place either as a one-off survey or a regular survey. These surveys provide spatial and temporal aspects of the travels and activities of the population sample at different granularities, which in turn help in understanding people's travel demand and their preferences. In contrast to a one-off survey, a regular survey reflects a general movement pattern and travel demand at different situations and various spatiotemporal granularities, whereas a one-off survey sometimes reflects biased travel behavior affected by circumstantial influence at a given time period. For instance, in 1973–1974, the French National Travel Survey, a one-off large scale survey, took place during the first oil crisis. In 1993–1994, the survey was again conducted during an extreme financial crunch. In both cases, owing to adverse socioeconomic situations, the numbers of trips were reduced compared to non-surveyed times and thus the overall travel data were biased, which provided misleading travel demand information (Ampt, Ortúzar, and Richardson 2009; Ortúzar et al. 2010). On the other hand, a regular survey, for example, the National Mobility Survey in the Netherlands, which is repeated every year, can capture travel behavior in different conditions (impact of seasonal variations and various events on people's travel demand) across the country.

There are various methods to design and conduct HTS (Zmud 2003). Table 8.1 presents the minimum required information collected in HTS suggested in the literature (Stopher et al. 2006). Besides, the information collected in HTS is extensive and detailed, resulting in various difficulties in the survey execution and inaccuracy in its outcome. Currently, a paper-based, face-to-face, or telephonic approach is used to conduct an HTS. Nevertheless, as the participants report about their travel at a later time, there is a chance that they may not remember all the details of their travels and activities, or may answer the questions with a bias. Moreover, they may grow disinterested

TABLE 8.1

Suggested Minimum Required Information Collected in HTS.

Category	Household	Personal	Vehicle	Activity
Items	Location	Gender	Body type	Start time
	Type of building	Year of birth	Year of production	Activity or purpose
	Household size	Paid jobs	Ownership of vehicle	Location
	Relationships	Job classification	Use of vehicle	Means of travel
	Number of vehicles	Driving license		Mode sequence
	Housing tenure	Non-mobility		Group size
	Re-contact	Education level		Group membership
		Disability		Costs
		Race		Parking
		Hispanic origin		

as the questionnaire is long and not answer all the questions; thus, it raises a gap in the data collection process (Wolf, Oliveira, and Thompson 2003; Stopher, FitzGerald, and Xu 2007).

8.2.2 GPS-Assisted Travel Surveys

With the introduction of advanced positioning technologies, such as GPS, the limitations of manual travel surveys have been mitigated through an automated approach. A GPS-assisted travel survey involves recording a person's travel behavior as a sequence of time-ordered spatiotemporal points, also known as a trajectory. The granularity of a trajectory can be affected by the predefined sampling interval and the GPS signal quality.

A proof-of-concept GPS-assisted survey was initially conducted in Lexington, Kentucky, as a part of a larger HTS (Battelle 1997; Murakami and Wagner 1999; Auld et al. 2009). GPS-assisted travel surveys can be categorized as in-vehicle based or handheld based. An in-vehicle approach uses a GPS receiver mounted on the vehicle, and thus reduces the burden on the participant's part. However, an in-vehicle GPS-assisted survey records only the vehicle travel history, and thus cannot record the portion of the trips travelled by walking, biking, or by public transport modes. Since the majority of urban activities take place while walking or being stationary in a constrained space, an in-vehicle GPS-assisted survey provides limited information on activity patterns of the participants.

In contrast to an in-vehicle GPS-assisted travel survey, a handheld GPS-assisted survey can generate activity-trip data across all modes (Chung and Shalaby 2005; Auld et al. 2009; Bohte and Maat 2009; Elango and Guensler 2010; Roorda, Shalaby, and Saneinejad 2011).

The participant has to carry the GPS logger along her travel and the device can record automatically.

The duration of such a survey can span from 1 day to several weeks depending on the survey requirements. However, the accuracy of the recorded data depends on where the GPS logger is kept and the number of satellites the receiver can view. Handheld GPS loggers are also subject to quick battery depletion. The participants have to always remember to carry the logger with them, which creates an extra mental burden on the survey participants. Owing to this stringent survey practice, participants may not follow their real travel behavior (Safi, Mesbah, and Ferreira 2013). This problem is further mitigated by using smartphones for travel data collection.

8.2.3 Smartphone-Based Travel Surveys

The recent advancement in the field of information communication technology (ICT), along with the positioning technology on one hand and the ubiquitous use of smartphones on the other, has facilitated the automation of HTS. This has been particularly possible due to the various sensors onboard the smartphones, for example, location sensors (GPS, GSM, and Wi-Fi), inertial measurement units (accelerometer, gyroscope, and compass), proximity sensors, and light and pressure sensors. Gonzalez et al. (2010) developed a smartphone-based travel survey application TRAC-IT in order to estimate people's travel demand in terms of their preferences and usage of different transport modes. The research involved 14 respondents. The application recorded positional information at four second intervals and transmitted the data to a data server immediately (Gonzalez, Hidalgo, and Barabasi 2008; Gonzalez et al. 2010). Charlton et al. (2010) developed CycleTracks—a smartphone-based application to understand cyclist's travel behavior. CycleTracks allowed participants to track their trips and upload them to a central server once the travel was complete. Similar applications have also been developed by other researchers with a focus on collecting either a person's activity patterns or travel behavior (Jariyasunant et al. 2011). However, most of them are limited in collecting fine-grained activity-trip data to support HTS.

Recently, the Singapore-MIT Alliance for Research and Technology (SMART) has developed a more sophisticated smartphone-based application, Future Mobility Sensing (FMS), as a part of Singapore HTS (Cottrill et al. 2013). FMS consists of a smartphone app, a server, a user interface, and a website (Raveau et al. 2016). Users of FMS install the app on their phone, open an account, and keep their smartphone charged. The sensors onboard the smartphone (GPS, GSM, accelerometer, and Wi-Fi) collect the users' movement data and transfer them to the FMS server, where the data are processed and various inferences about the mobility of the users are drawn (such as travel time, travel distance, and mode). These automatic inferences are presented to the users on their private FMS website in the form of a daily timeline for their review and validation (Raveau et al. 2016). Similar surveys have also been developed and tested elsewhere (Safi, Mesbah, and Ferreira 2013).

HTS (automated and non-automated) belongs to the category of revealed preference (RP) surveys that reveal the current state of people's travel behavior and preferences, hence the name. The main advantage of RP surveys is that they represent the actual users' choices reflecting all their real-world constraints and perceptions. As opposed to RP surveys, there are stated preference (SP) surveys that investigate the users' behavior in mostly hypothetical scenarios. Both surveys have their own specific use cases and applications. It is also common to conduct studies using a mixture of both surveys to enrich the datasets (for more information, see Hensher, Rose, and Greene (2005)).

8.2.4 Data Collected from Car Navigation Systems and Other Means

Instead of surveys, and pushing the boundaries of smartphone-based information collection, it also has become common practice to collect mobility data through navigation or mobility applications with significant sampling rates if not the full population. Car navigation systems, both stand-alone or on smartphones, track the movements of cars in high temporal resolution, mostly relying on satellite positioning systems, but also integrating inertial sensors. These kinds of data are immediately used by Web mapping platforms to predict current traffic conditions in an aggregate form. Thus, these new sources of data are unprecedented in terms of volume, geometric accuracy, temporal resolution, and thematic variation (Li et al. 2016).

The collected data from these new sources are also used for travel demand prediction, both at the aggregate level (anticipation of traffic conditions [Sevtsuk and Ratti 2010]) and individual level (e.g., detecting joint demand [Santi et al. 2014]). Although limited to a particular mode of traveling, the driver only, by crossmatching these data to land use data, it is possible to infer the demand based on a significant share of the population, not just a small sample.

Navigation systems for other modes of transport collect similar data, for example, mobile visitor guides. If they operate indoor, that is, satellites signal-deprived environments, such as museums, they use different positioning technologies. However, their tracking data are analyzed for the same interest of travel demand analysis, although at this fine scale and high spatial granularity (Yalowitz and Bronnenkant 2009).

Next to tracking by smartphones, web applications such as public transport planners or the taxi, train, or air travel booking systems also collect data (from queries or from bookings) that can be analyzed for travel demand. Furthermore, in the cities where smartcards are used as public transport tickets, the public transport usage and travel demand of the full population are tracked (Zhong et al. 2016).

The accuracy and scale of the data acquired from these types of data sources may revolutionize the field of demand modeling, as they have revolutionized the transport supply model. In the supply field, it has facilitated the deployment of demand responsive transport modes, such as the recent ridesharing platforms, the provision of real-time information to public transport users, or the provision of network traffic to private vehicle users (Miller and Shaw 2015). Nonetheless, the restrictions of each of these datasets are significant. All of them are sectoral, being limited to a particular travel mode or even a particular transport provider. For reasons of commercial sensitivity as much as privacy of the users, these data are not accessible outside these companies. Therefore, no integrated transport demand can be produced from these data yet. It is only used within the companies for adapting their own services to the derived demand.

8.3 Inferring Activity-Trip Information from Trajectory Data

The raw trajectories collected by a smartcard, or GPS sensors installed on a vehicle or on a smartphone, provide only the traveler's location information at a given time period (Figure 8.1). However, a transport planner wants to know more about the user's travel behavior, such as the transport mode choice, origin, destination, route taken, transfers, accompaniments, and activity types

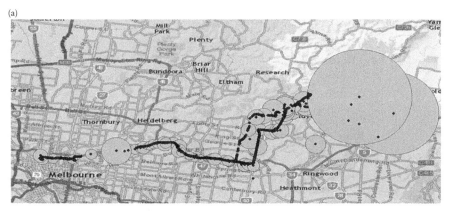

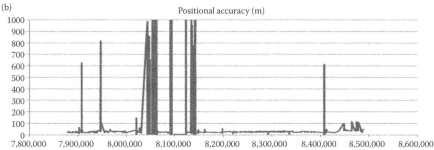

FIGURE 8.1
A raw trajectory (locations visited) on a map (a) and as a space-time graph reflecting how time elapsed during travel (b).

at an individual as well as aggregate level. This latter information (i.e., the semantics) cannot be conceived directly from the raw trajectories. Hence, the raw trajectories need to be analyzed to extract the semantics pertinent to travel demand modeling. Before any analysis takes place, a raw trajectory is generally preprocessed to remove any noise (generally occurring due to multipath effect in urban canyons) based on positional accuracy information or the number of visible satellites (Xiao, Juan, and Gao 2015). An in-house study has shown that when a trajectory is recorded by different smartphone sensors, such as GPS and GSM, the accuracy level varies significantly from a few meters to thousands of meters (Figure 8.2), depending on the GPS signal reception and cell tower distribution in the study area.

A preprocessing procedure can also resample a raw trajectory in order to remove noisy GPS points through various interpolation techniques (Long 2016) or kinematic measures (Stenneth et al. 2012, Lari and Golroo 2015). A preprocessing operation also involves time conversion if required, and data projection from one coordinate system to another for further spatial computation (Wu, Yang, and Jing 2016).

Depending on the information needs, the preprocessed trajectories can be semantically enriched by incorporating different contextual information as follows:

1. *Spatial information*: This type of information includes route network, point of interest (POI), or region of interest (ROI) data.

(a)   (b)

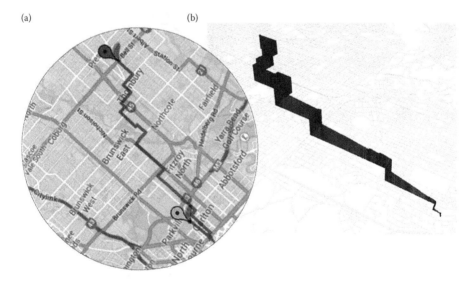

FIGURE 8.2
The accuracy (a bigger ellipse means less spatial accuracy) of a raw trajectory recorded by several positioning sensors (a), and accuracy levels (Y) with respect to spatiotemporal points (X) (b).

2. *Spatiotemporal information*: This includes general transit feed specification (GTFS), or opening and closing hours of various facilities.

3. *Thematic information*: This type of information can be expressed, for example, by land use data, or event type data.

4. *Socioeconomic information*: This contextual information includes the social status of people and their interactions with their environment, the number of people travelling along a route or on a mode, or the income profile of people.

In terms of response time required for a particular application domain, a preprocessed trajectory can be analyzed both offline, once the travel is complete, or online, during the travel itself.

An offline-based inference generates historical movement patterns for long-term travel demand modeling and policy enforcements, whereas a real-time trajectory inference helps in ad hoc demand estimation and real-time travel information. In order to analyze raw trajectories offline, the majority of researchers use a top-down trajectory segmentation approach, which breaks a trajectory into a number of homogeneous segments. Then the analysis takes place over each segment, and an activity state is detected over the given segment. Spaccapietra and Parent (2011) developed such an episodic algorithm for trajectory segmentation known as stop-and-move-on-trajectories (SMoT) (Bogorny et al. 2011; Spaccapietra and Parent 2011). The basic assumption behind this algorithm is that a person will stop at a given location for a certain time period in order to conduct an activity and then start travelling to her next destination. That said, in a SMoT algorithm, a move episode reflects a person's travel behavior, whereas a stop episode reveals a person's activity behavior within a constrained space. There are three different variants of SMoT algorithms as follows:

1. *Intersection-based stop-and-move (IB-SMoT)*: This approach initially finds the spatiotemporal points in a raw trajectory that are within a given candidate region over a certain time duration in order to detect the stop episodes, and the remaining points will be inferred as move episodes (Alvares et al. 2007).

2. *Clustering-based stop-and-move (CB-SMoT)*: In the second variant of SMoT, a clustering is performed to detect the stop points based on their spatial proximity and low speed profiles. The points that do not fall in the clusters are classified as move points (Palma et al. 2008).

3. *Direction-based stop-and-move (DB-SMoT)*: In the third variant, the spatiotemporal points in a trajectory are classified as stop points or move points based on the change in directions (Rocha et al. 2010).

Similar to a CB-SMoT approach, a number of researchers have developed clustering-based segmentation approaches independently (Ashbrook and

Starner 2003; Zimmermann, Kirste, and Spiliopoulou 2009; Gong et al. 2015). Since walking is required between any two motorized (or bike) modes of transport, Zheng et al. (2008) proposed a speed- and distance-based segmentation approach where a raw trajectory is first segmented into a number of walking and non-walking segments. An interpretation process is then performed to infer the specific activity over the given segment. In addition to a distance- or speed-based measure, the temporal aspect can also be considered in order to segment a trajectory (Andrienko et al. 2013).

In order to detect an activity from a GPS trajectory, the existing approaches are mainly rule based. An activity is identified if there is a longer period of non-movement, or longer dwell time at a certain location. An activity can be a trip end, going home or office, meeting friends at a restaurant, having coffee at a cafeteria, or transferring from one transport mode to another. Prior studies suggested that if the dwell time is greater than 120 seconds, then that could be a probable trip end (Stopher 2004; Clifford, Zhang, and Stopher 2008; Bohte and Maat 2009). However, there might be cases of shorter stops than 120 seconds, for example, when a car stops for passengers to get on or drop off the vehicle while the engine is still running. To address such diverse activities, several deterministic dwell time algorithms have been devised. For example, Wolf et al. (2004) developed a hierarchical approach to detect trip ends with varied confidence proportional to the amount of dwell time at a given location.

In order to characterize a trip purpose, Wolf et al. (2004) developed a point-in-polygon approach, which was based on 25 predefined land use types and 11 trip purpose classes. A point-in-polygon approach first detects the trip start and end points. Then the trip end points are spatially correlated to the nearest polygon that bears a specific land use type and trip purpose type. However, a point-in-polygon approach requires a comprehensive and accurate GIS database containing network and land use information. The approach fails when there is a GPS signal gap or multipath effect. To address this issue, Stopher et al. (2008) have proposed a rule-based method that used trip characteristics before and after the GPS signal loss. Stopher, Bullock, and Jiang (2002) also investigated how the accuracy varies on incorporating GIS information. On the other hand, in order to address the multipath effect, Wolf et al. (2004) proposed a clustering algorithm to detect the trip end points from a number of trajectories. Within the clusters, all the POIs are spatially queried within a given search radius (e.g., 300 m) of the cluster center. Once the relevant POIs are retrieved, each of them is assigned a weight based on their proximity to the cluster center. The higher the weight, the more likely the given POI is the trip destination. Along the same lines, some researchers have developed similar types of rule-based algorithms to detect the trip destination and trip purpose. For example, based on the frequency of visits, the spatial proximity to a given POI, and the affordance offered by that POI, the trip purpose has been inferred from trajectories (Wolf et al. 2004; Clifford, Zhang, and Stopher 2008; Schuessler and Axhausen 2009).

Since rule-based and deterministic-activity recognition models are not able to address dynamic activities with various dwell time and spatial constraints, machine learning and probabilistic approaches have been explored to improve the recognition process. Liao, Fox, and Kautz (2005) used conditional random fields and relational Markov networks to detect and rank various activities. In order to calculate the trip lengths, two common approaches exist currently: a point-to-point (PP) approach and a link-to-link (LL) approach (Murakami and Wagner 1999). In the PP approach, distance is calculated between two consecutive points over a preprocessed trajectory, and then the distance is cumulatively added up over an entire segment or a trip. However, this approach is subject to GPS positional uncertainties, especially in the urban canyons or indoor environments. To address this issue, a route network is considered where a portion of the trajectory is compared with a given segment of the route network using a map-matching algorithm. This way, even if there is a signal gap, if trip origin and destination and the route are detected properly, the length of the trip can be computed using the route network.

As discussed earlier in this section, trip characterization is an important aspect of transport demand modeling to generate the origin–destination (OD) matrix. However, understanding the mediation of transport is also critical for estimating the patronage information and people's mode choice behaviors. Transport modes can be detected either offline (on historical trajectories) or online (in real time). Although there is no clear guideline between an offline and online trajectory interpretation process, an offline-based mode detection approach involves two stages of operations. In the first stage, a segmentation is performed, followed by detecting a particular transport mode over given segments by a machine learning approach, a rule-based technique, or a hybrid approach. In order to execute a prediction algorithm, a number of features are computed over each segment. For example, Zheng et al. (2008) considered four modalities: walk, car, bus, and bike. They computed a number of kinematic features, such as mean velocity, heading rate change, and top three acceleration values to test four different machine learning models, a decision tree, a Bayesian network, a conditional random field, and support vector machines. Zheng and colleagues obtained the maximum prediction accuracy (76%) using a decision tree-based model. Similarly, Stenneth et al. (2011) classified six modalities (car, bus, train, bike, stationary, and walk) using six different machine learning models such as Bayesian network, decision tree, random forest, and a naïve Bayes, with accuracies of 92.5%, 92.2%, 93.7%, and 91.6%, respectively, by incorporating infrastructure information. Gonzalez et al. (2010) computed a number of features such as average acceleration, average speed, and maximum speed, ratio of critical points over a trip distance, and duration of travel for their neural network-based critical point model in order to get rid of voluminous GPS data while performing spatial computation. They detected three modalities: car, bus, and walk. They have demonstrated that the prediction accuracy increases when spatial information is added. On the other hand, Dodge, Weibel, and Forootan (2009) have introduced the concept of local and global features while computing various kinematic and non-kinematic features over segments. They considered variation in sinuosity and deviation of different kinematic features. Xiao et al. (2017) used the concept of global and local features while evaluating three different machine learning models: random forest, gradient boosting decision tree, and XGBoost.

In order to detect transport modes in real time, a temporal kernel is run over a trajectory, and a number of features are computed within that kernel. These computed features are then used to train or test a predictive model that can retrieve a transport mode information at a given time period. In this context, the model developed by Reddy et al. (2010) is relevant where they used a GPS and accelerometer to compute speed measures- and acceleration-based features. Their model shows 74% accuracy. Hemminki, Nurmi, and Tarkoma (2013) used a discrete hidden Markov model along with AdaBoost while detecting different modes with 84.2% accuracy.

Since existing trajectory interpretation models either work online or offline, a given model cannot adapt to different response times required to generate travel information. To address this need, Das and Winter (2016a) proposed a more sophisticated and adaptive trajectory interpretation model that can adapt to different contexts and provide travel information at different temporal granularities. The model introduces a bottom-up trajectory segmentation approach that assumes a trip is an aggregation of short atomic segments of homogeneous modal states. Thus, by merging the homogeneous states iteratively, a trip can be detected over a given time period. In order to raise the trust in an inference process, a number of lemmata have been proposed. Unlike earlier trajectory interpretation models that use either spatial information or temporal information, the bottom-up model uses spatiotemporal information (GTFS) while detecting a transport mode along a given route.

Although machine learning models work effectively, the models require a significant amount of data to get trained. Thus, a machine learning model falls short in an environment where the training data are limited. A machine learning model also provides limited explanatory ability while interpreting a trajectory. To address these limitations, rule-based frameworks, in particular fuzzy logic-based models, have been proposed. A fuzzy logic model is based on expert knowledge and does not require any training data. Thus, a fuzzy logic-based model allows several domain experts to interact with the model and develop the rule base in a customized way, which is not the case for a machine learning-based model. In transport mode detection, Schuessler and Axhausen (2009) proposed a fuzzy logic-based approach that can distinguish five transport modes using speed-only features. Xu et al. (2010) presented a similar type of fuzzy logic-based model that can detect four modalities with 94% accuracy. Biljecki, Ledoux, and Oosterom (2013) developed a more comprehensive hierarchical fuzzy logic-based knowledge driven model that can distinguish ten different transport modes with 92% accuracy. Although fuzzy logic-based models can express their reasoning scheme, the models lack the adaptivity of machine learning. In order to bridge the trade-off between adaptivity and explanatory ability, a hybrid model has been proposed by integrating a neural network and fuzzy logic to detect different transport modes based on their speed profile and proximity to different route networks (Das and Winter 2016b).

By changing the focus from a disaggregated perspective to an aggregate perspective, a collective inference can be performed on a number of trajectories generated by different users. Unlike an individual trajectory interpretation, a collective trajectory interpretation can generate aggregate demand information in terms of hot spots (Gudmundsson, Kreveld, and Staals 2013), top-k routes during peak hours, or urban form and functions and their evolutions over time (Crooks et al. 2015). With the emergence of user-generated contents (UGC), there is a rapid proliferation of social media data across the globe. Social media data are deemed to provide additional potential for studying travel demand at an aggregate level. Lee et al. investigated the feasibility of Twitter data for travel demand estimation in comparison to a manual HTS in California. The results demonstrate that Twitter data can generate similar travel information as that of a manual HTS, especially in terms of trip length and spatial distribution of trips (Lee et al. 2017). However, the amount and quality of tweets largely depends on the sociodemographic profiles of the users of social media, the type of transport modes used during travel, and the users' perception about an event and the way they express their reactions. Similar approaches have also been explored by analyzing geo-tagged photos extracted from platforms such as Flickr or Foursquare (Zheng, Zha, and Chua 2012). Although such platforms may not provide detailed travel information all the time, the coarse-grained information can be useful for understanding travel demands in the interest of the tourism industry or location-based e-marketing services.

In this section, a wide range of geospatial data sources and their applications in travel demand modeling have been explored. Unlike a manual travel survey, where people report about their activity-travel patterns from their past memories, smartphone-based dedicated tracking and UGC datasets provide not only people's activity-travel information but also people's interaction with a place or an urban event and how that interaction can impact travel demands over a shorter or longer term. There is also a possibility to combine different datasets (spatial, temporal, thematic, and socioeconomic), leading to big data analytics for transport planning. However, with the real-time tracking along with the growing variety and volume of movement data, there is an open challenge as how to store, manage, and extract travel information from big data. Besides, geospatial mobility datasets raise privacy concerns, which limit the acceptance of these techniques as long as no privacy aware handling of the data is established. Current research has started looking into these issues from the perspective of human geography, transportation engineering, and data science for effective transport demand modeling.

8.4 Demand Modeling

Transportation demand comes from the spatial differences of origins and destinations of people's activities, that is, working, leisure, and shopping, and grows constantly due to the urbanization phenomena and the expansion of cities. Understanding and forecasting this ever-growing demand plays a crucial role in making efficient decisions for transportation infrastructure and policy, which is only possible through what is known as demand modeling.

Demand modeling was started in the 1950s in the United States with the four-step model (FSM), which consists of the following steps:

1. *Trip generation*: The main purpose of this first step is to identify the number of trips starting and ending in a specific zone based on socioeconomic characteristics of the users residing in that zone as well as the zone's land use data.

2. *Trip distribution*: In the second step, the origin and destination of each trip are identified based on the number of produced and attracted trips (i.e., the previous step's outcome) from a zone and the impedance value between two zones, which depends on the time and effort required to travel from one zone to another.

3. *Modal split*: The third step is to determine the mode of each trip based on the modal split data acquired from surveys (e.g., HTS), which could become very complex as the variety of modes increases.

4. *Trip assignment*: The last step is dedicated to assigning a route on the transportation network to each trip.

Although FSM represents the travel behavior, it has failed to reflect the main underlying reason for traveling, that is, performing activities (McNally and Rindt 2008). Another main criticism of FSM is the lack of feedback in the sequential procedure of FSM. For instance, the number of trips is determined in the first step regardless of the transportation services. In the real world, if a person has difficulties accessing a location to perform an activity, he/she might avoid traveling in the first place, affecting the number of trips. In spite of the aforementioned issues, FSM has been widely used as the main demand forecasting method for many projects even to this date, particularly for developing large-scale infrastructures.

Nonetheless, the beginning of disputes about environmental issues, space and budget limitation, and traffic congestion in the 1970s challenged the unconditional development of infrastructure and called for a more astute method for demand modeling that is able to effectively reflect the impact of different policies and management strategies to increase the efficient utilization of the existing infrastructure. Contemporaneously, three independent studies established the foundation of a notion that was later called activity-based approach (ABA) for demand modeling (Miller and Shaw 2015). Hägerstrand (1970) introduced the spatiotemporal constraints of activity participation; Chapin (1974) investigated and described people's behavioral pattern across space-time; and Fried et al. (1977) examined people's activity participation motives from a societal point of view. However, the connection of travel, activity, and space-time was explicitly addressed for the first time by Jones (1979) and began to receive considerable attention after a conference held in 1981 on "Travel demand analysis: activity-based and other new approaches" (see Carpenter and Jones 1983 for proceeding).

Unlike FSM that has a clear and slightly strict structure, ABA is more of a concept or theory rather than methodology mostly due to the complexity of the considered factors, such as individuals' characteristics, their relation to their household members' travel patterns, spatial and temporal aspects of activities, and other social circumstances (McNally and Rindt 2008). Notwithstanding the diversity of applied methodologies and empirical approaches, they all fall within four categories: simulation-based models, computational-process models, econometric-based application, and mathematical programming approaches (see McNally and Rindt (2008) and Pinjari and Bhat (2011) for detailed descriptions), and share one or more of the following characteristics (McNally and Rindt 2008):

- Travel is a result of demand for activity participation;
- The analysis units are sequences or patterns of behavior (not individual trips);
- Social structures, on any level, influence travel and activity pattern;
- Spatial, temporal, transportation, and interpersonal interdependencies constrain both activity and travel behavior; and
- ABAs reflect the temporal and spatial aspects of activities planning.

Considering the above features, it can be concluded that a highly detailed level of information about individuals, their socioeconomic characteristics, and their activities is required. Although similar data are collected in HTSs (see Section 8.2), it is not possible to directly develop a population-wide demand model on an individual level due to two main reasons. First, the data are just acquired for a limited number of individuals in each study area, that is, a sample of the population. Second, owing to privacy issues, the origins and destinations of the trips are reported in an aggregated level: not as coordinates but area codes. Additionally, depending on the HTS that the data are derived from, the start time or duration of the activities may also be unknown.

To tackle the first problem, it is possible to generate a synthetic population based on the sample data. The precision of the synthetic population greatly depends on the employed approach and algorithm (Lim and Gargett 2013).

There are several approaches that can be utilized to generate synthetic population, including: stratified sampling, geodemographic profiling, data fusion, data merging, iterative proportional fitting, reweighting, synthetic reconstruction, and their combinations (Huang and Williamson 2001). Synthetic reconstruction (SR) and combinatorial optimization (CO), which is an alteration of the reweighting approach, are two major approaches (Wilson and Pownall 1976; Williamson, Birkin, and Rees 1998; Williamson 2002). Using different methods, both approaches create individuals and households with consistent characteristics to the known aggregate distributions of the census (see Ryan, Moah, and Kanaroglou 2009; Huynh et al. 2013; Lim and Gargett 2013; Jain, Ronald, and Winter 2015 for more information). Several software applications have been developed over the years based on the mentioned approaches, namely PopSynWin (Auld, Mohammadian, and Wies 2008) and PopGen (Ye et al. 2009), which have been effectively utilized to generate the micro data in different cities. Since the synthetic population is an extrapolation of the HTS, it just solves one of its shortcomings to generate a population-wide daylong tour-based demand.

To address the second mentioned obstacle, that is, the lack of specific spatial and temporal features of the activities, researchers employ different heuristic approaches, all of which share the same challenges: integration of land use data, and activities' time and location generation and allocation. While the latter directly addresses the specifications of trips, the former is necessary to ascertain that the trips' origins and destinations are valid not only in terms of the reported area but also in terms of land use data. For example, measures need to be taken that a trip from home in zone 1 to a store in zone 2 starts from a point in a residential area in zone 1 to a commercial area in zone 2. To that end, transport modelers and researchers exploit various geospatial databases and software packages and combine them with transportation concepts and techniques.

Bowman et al. (1998) and Bhat, Srinivasan, and Guo (2001) first utilized households, zonal data, and network data to create a tour-based travel plan for individuals and generated alternative origin and destination locations. Then they applied econometric methods to allocate time and location to each activity in tours. Figure 8.3 demonstrates the overall work structure of Bowman et al. (1998) in Portland.

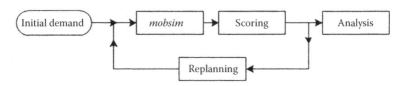

FIGURE 8.3
Portland activity schedule model system. (Modified from Bowman, J.L. et al. 1998. Demonstration of an activity based model system for Portland. *Paper presented at 8th World Conference on Transport Research.* Antwerp, The Netherlands.)

Rieser et al. (2007) took a simulation-based approach to generate the population-wide demand for Berlin. They first created initial plans, that is, a sequential list of activities' location and time, and then used a transport mode between them, for all individuals based on the Kutter Model (Kutter 1984; Kutter and Mikota 1990; Kutter et al. 2002); then they used it in a multiagent simulation software to replicate the real-time and location choice of people. Their model was validated against traffic counts from 100 stations.

8.5 Simulation

Transport models or, more specifically, transport demand models are necessarily simplified representations of a complex and dynamic transportation system. Transport models describe and display a static state of the system (with limited complexity). To investigate the changes to the state of a system, and to understand the effect of these changes, simulations are necessary. Furthermore, simulations can be utilized to synthesize data or create the demand model in many cases. For instance, Hensher, Rose and Greene (2005) proposed data synthesis through simulation as a solution to the missing non-chosen alternatives RP data problem. Also, agent-based simulation is recommended as one of the techniques to create the demand model in ABA (see Section 8.3).

The simulation of transport systems started in the 1950s with car-following behavior analysis in a platoon, which facilitated the study of traffic flow (Gerlough and Huber 1975). Since then, simulations have been in use to investigate various aspects of traffic systems, such as flow, traffic signal control, and intersections (Persula 1999). The most common classification system for the classic traffic/transport simulations is based on the degree of the details they can consider. The literature suggests three levels of classic traffic/transport simulation (Fellendorf and Vortisch 2010):

- Simulation at a *macroscopic* level sees the traffic as a fluid and characterizes it (macroscopically) by volume, density, and speed;
- Simulation at a *microscopic* level describes the movement and characteristics of every individual vehicle (such as its position, speed, acceleration, and lane changes), considering all the surrounding vehicles and the environment;
- Simulation at a *mesoscopic* level is a mixture of the former ones. In a mesoscopic simulation, each vehicle moves with the macroscopic quantities and the results are referring to the individual vehicles.

The advancements in a number of technological and scientific fields, including computer hardware, computer science, data science, complexity studies, and the global positioning system (GPS), lead to the introduction of a new category of transport microsimulation that is focused on individuals rather than vehicles, that is, geosimulation (Torrens 2004; Benenson and Torrens 2004). This innovative class of transport microsimulation specifically facilitates the study of travel demand on an individual person's level, which allows the exploration of policies to better utilize the existing infrastructure and improve the social inclusion and justice in the mobility sector.

Benenson and Torrens (2004) refer to geosimulation's capability to represent the components of a transport system as different objects with specific characteristics and relations as its main feature that differentiate it from traditional approaches in transport or urban system simulation and describe its four core characteristics as follows:

- *Representation of spatial entities*: Unlike the traditional approaches, geosimulation can represent geographic entities on a spatially non-modifiable level, such as households, home, and vehicles.
- *Representation of relationships*: While in traditional approaches the interactions happen on an aggregated level, for instance among traffic analysis zones or cities, in geosimulation the synergy of urban entities forms the interaction at higher levels.
- *Treatment of time*: In geosimulation, different objects can behave in a different temporal manner: synchronous, concurrent changes in all objects, or asynchronous, when changes happen in turn.
- *Direct modeling*: Geosimulation's potential to model the real behavior of the objects on a fine level and realistic manner enables the researchers to develop the proper scientific tool for scrutinizing hypothesis and theories (Benenson and Torrens 2004) in a way that was not possible previously.

An automaton, "… a machine that processes information, proceeding logically, inexorably performing its next action after applying data received from outside itself in light of instructions programmed within itself" (Levy 1992, 15), is the means to implement the object-oriented concept of geosimulation. Mathematically, an automaton is defined as:

$$S_{t+1} = f(S_t, I_t)$$

where S is the state of an automaton at a given time, f is the transition rule for the automaton from time t to time $t + 1$, and I represents the input data from outside of the automaton, that is, from neighboring automata.

Benenson and Torrens (2004) describe the representations of the traffic system elements in an automaton-based scheme: automata can represent any component of a traffic system, such as vehicles, pedestrians, and sections of road. These elements' state can be described with different attributes, such as capacity, speed limit, and demographic characteristics. Input data can be inferred from transition rules that are defined to represent the behavior and processes, such as lane-changing rules, rules describing people's preferences or motions. All these automata can be integrated into different types of lattices representing the spatial context of a transportation system, such as regular grid-based tessellations, irregular grids, and graph-based networks of nodes and edges, with various time scales.

The concept of automata is applied using Cellular Automata (CA) or multiagent systems (MAS). In CA, the automata are bound to move within a cellular, for example, grid square, environment with a limited number of neighbors and interactions, which results in a limited number of CA applications in transport systems. In contrast, multiagent systems are more flexible in terms of the definition of the environment, the agents' neighbors can be positioned randomly around them, and the agents can have "arbitrary neighborhood connections" (Benenson and Torrens 2004).

MAS have become the basis for developing numerous simulation software packages for transport studies, namely Transportation Analysis and Simulation System (TRANSIMS) (Smith, Beckman, and Baggerly 1995), Multiagent Transport Simulation (MATSim) (Horni, Nagel, and Axhausen 2016), and Comprehensive Econometric Microsimulator for Activity-Travel Patterns (CEMDAP) (Bhat et al. 2004; Pinjari et al. 2006).

Understanding the notion and features of MAS confirms its effective application in ABA. An agent in an MAS can be defined as "… a system situated within and a part of environment that senses that environment and acts on it, over time, in pursuit of its own agenda, and so as to effect what it senses in the future" (Franklin and Graesser 1996, 25). In other words, it is an autonomous entity that has purpose, preferences, and a perception of its environment, and acts based on those inputs to impact its future.

For instance, MATSim is an established multiagent simulation software for modeling and simulating transport phenomena that integrates ABA into an agent-based simulation system. MATSim is employed for large-scale simulations of real-world scenarios. The overall workflow of MATSim is illustrated in Figure 8.4.

The *initial demand* includes a list of all people, represented by agents, and their daily plans including the locations and times of their daily activities, which are generated based on a relevant HTS (see Section 8.3). During the *mobsim* phase, all the agents execute their plans in a predefined environment, which includes transport network and various facilities' properties (optional), and save the attributes of their executed plan, for example, travel time, waiting time at a stop, waiting time for a workplace or shopping center to open, or arrival time. In other words, agents travel through the network to reach their activities' location and, depending on the other agents' plan, they may get stuck in a traffic jam or arrive earlier than they expect. In the next step, *scoring*, the executed plans' score is calculated based on the saved attributes using scoring functions that are defined generally for all agents or

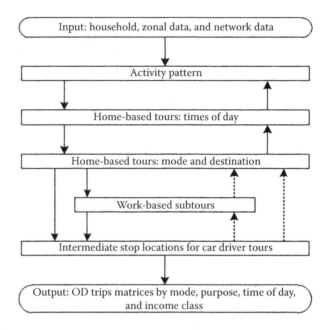

FIGURE 8.4
The overall workflow of a MATSim loop (or cycle). (Modified from Horni, A et al. 2016. *The Multi-Agent Transport Simulation MATSim.* Ubiquity, London.)

individually for each agent. In the *replanning* step, a number of agents make changes in their plan and the system is ready for another cycle. Each agent has a limited number of memories for saving plans and after each scoring phase, it updates its memories to contain the most suitable plans. This loop is repeated in several iterations until it reaches an equilibrium state, where no one can significantly improve their plans anymore (Horni, Nagel, and Axhausen 2016). Integration of Belief-Decision-Intention platform as a recent extension to MATSim even allows for a smarter interaction with the environment by provoking a decision-making process at any potential decision point during the *mobsim* phase (Padgham et al. 2014). Table 8.2 presents a comparison of the features of a general agent to those of an agent in a transport simulation system.

TABLE 8.2

Comparison of a General Agent to a MATSim Agent.

General Agent Definition	MATSim Agent
System situated within and a part of the environment	Transport network and various facilities' properties
Senses that environment and acts on it, over time	In the *mobsim* phase, they interact with their environment and other agents
In pursuit of its own agenda	All agents represent people, who have daily plans

Many of the agent-based simulation software with a similar structure have compromised modeling the details of transportation systems (such as exact location or speed of vehicles at each time step, and driving behavior) for modeling urban phenomena, such as crime, segregation, and disease within cities (Wise, Crooks, and Batty 2016).

8.6 Conclusion

Transportation demand is the result of people's needs and desires to perform activities in spatially dispersed locations. Understanding and modeling this phenomenon is the key to understanding the transport demand and solving many urban problems related to transportation and land use. ABA has been tested and proven to be the prevalent approach capable of reflecting this complex intertwined relation of people, activity, and travel, and requires data on the finest level possible, which is socioeconomic characteristics of people, potential locations for different activities, as well as possible and desired time to perform activities. This chapter provides an overview of different geospatial data sources and some commonly used inference strategies to extract activity-travel information from those geospatial data sources. In this regard, it can be stated that geospatial data, collected through surveys or various automated sources (e.g., smartphones and smart cards) and UGC, have been revealed as the ultimate solution to the data hungry nature of ABA. Furthermore, geosimulation, particularly the agent-based simulation method and its object-oriented characteristics, paved the path for implementation of this complex and comprehensive theory. In the era of big data, where geospatial data come from different sources with different qualities and in huge volume, there is a future research direction that should look into combining and processing such complex datasets to extract relevant information and ensure a sustainable urban mobility.

References

Alvares, L. O., V. Bogorny, B. Kujipers, J. Macedo, B. Moelans, and A. Vaisman. 2007. A model for enriching trajectories with semantic geographical information. In *Proceedings of the 15th Annual ACM International Symposium on Advances in Geographic Information Systems*, 22. Seattle, WA.

Ampt, E. S., J. de D. Ortúzar, and A. J. Richardson. 2009. Large-scale ongoing mobility surveys: The state of practice. In *Transport Survey Methods: Keeping up with a Changing World*, edited by P. Bonnel, M. Lee-Gosselin, J. L. Madre, J. Zmud, 503–531. Bingley, UK: Emerald Group Publishing Limited.

Andrienko, G., N. V. Andrienko, G. Fuchs, A. M. Olteanu Raimond, J. Symanzik, and C. Ziemlicki. 2013. Extracting Semantics of individual places from movement data by analyzing temporal patterns of visits. In *Proceedings of the First ACM SIGSPATIAL International Workshop on Computational Models of Place*, 9–15. Orlando, FL.

Ashbrook, D. and T. Starner. 2003. Using GPS to learn significant locations and predict movement across multiple users. *Personal and Ubiquitous Computing*, 7(5): 275–286. doi: 10.1007/s00779-003-0240-0.

Auld, J., A. K. Mohammadian, and K. Wies. 2008. Population synthesis with control category optimization. Paper presented at *10th International Conference on Application of Advanced Technologies in Transportation*. Athens, Greece.

Auld, J., C. A. Williams, A. Mohammadian, and P. C. Nelson. 2009. An automated GPS-based prompted recall survey with learning algorithms. *Transportation Letters: The International Journal of Transportation Research*, 1(1): 59–79.

Battelle Transportation Division. 1997. *Lexington Area Travel Data Collection Test*. Final report to Federal Highway Administration, U.S. Department of Transportation, Washington, DC.

Benenson, I. and P. M. Torrens. 2004. *Geosimulation: Automata-Based Modeling of Urban Phenomena*. Hoboken, NJ: John Wiley & Sons.

Bhat, C. R., J. Guo, S. Srinivasan, and A. Sivakumar. 2004. Comprehensive econometric microsimulator for daily activity-travel patterns. *Transportation Research Record: Journal of the Transportation Research Board* 1894: 57–66.

Bhat, C. R., S. Srinivasan, and J. Guo. 2001. *Activity-based travel demand modeling for metropolitan areas in Texas: Model components and mathematical formulations*. No. FHWA/TX-0-4080-2.

Biljecki, F. 2016. A scientometric analysis of selected GIScience journals. *International Journal of Geographical Information Science* 30(7): 1302–1335. doi: 10.1080/13658816.2015.1130831.

Biljecki, F., H. Ledoux, and P. Van Oosterom. 2013. Transportation mode-based segmentation and classification of movement trajectories. *International Journal of Geographical Information Science* 27(2): 358–407.

Bogorny, V., H. Avancini, B. C. de Paula, C. R. Kuplich, and L. O. Alvares. 2011. Weka-STPM: A software architecture and prototype for semantic trajectory data mining and visualization. *Transactions in GIS* 15(2): 227–248. doi: 10.1111/j.1467-9671.2011.01246.x.

Bohte, W. and K. Maat. 2009. Deriving and validating trip purposes and travel modes for multi-day GPS-based travel surveys: A large scale application in the Netherlands. *Transportation Research Part C: Emerging Technologies* 17: 285–297.

Bowman, J. L., M. Bradley, Y. Shiftan, T. K. Lawton, and M. Ben-Akiva. 1998. Demonstration of an activity based model system for Portland. Paper presented at *8th World Conference on Transport Research*. Antwerp, The Netherlands.

Carpenter, S. and P. M. Jones. 1983. *Recent Advances in Travel Demand Analysis*. Aldershot, UK: Gower.

Chapin, F. S. Jr. 1974. *Human Activity Patterns in the City: Things People Do in Time and Space*. London: John Wiley and Son.

Charlton, B., M. Schwartz, M. Paul, E. Sall, and J. Hood. 2010. CycleTracks: A bicycle route choice data collection application for GPS-enabled smartphones. Paper presented at *3rd Conference on Innovations in Travel Modeling, a Transportation Research Board Conference*, Tempe, USA.

Chung, E. H. and A. Shalaby. 2005. A trip reconstruction tool for GPS-based personal travel surveys. *Transportation Planning and Technology* 28(5): 381–401.

Clifford, E. J., J. Zhang, and P. R. Stopher. 2008. Determining trip information using GPS data. Working paper, ITLS-WP-08-01. Institute of Transport and Logistics Studies, The University of Sydney.

Cottrill, C., F. Pereira, F. Zhao, I. Dias, H. Lim, M. Ben-Akiva, and P. C. Zegras. 2013. Future Mobility Survey: Experience in developing a smartphone-based travel survey in Singapore. *Transportation Research Record: Journal of the Transportation Research Board* 2354: 59–67.

Crooks, A., D. Pfoser, A. Jenkins, A. Croitoru, A. Stefanidis, D. Smith, S. Karagiorgou, A. Efentakis, and G. Lamprianidis. 2015. Crowdsourcing urban form and function. *International Journal of Geographical Information Science* 29(5): 720–741.

Das, R. D. and S. Winter. 2016a. Automated urban travel interpretation: A bottom-up approach for trajectory segmentation. *Sensors* 16(11): 1962.

Das, R. D. and S. Winter. 2016b. Detecting urban transport modes using a hybrid knowledge driven framework from GPS trajectory. *ISPRS International Journal of Geo-Information* 5(11): 207. doi: 10.3390/ijgi5110207.

Dodge, S., R. Weibel, and E. Forootan. 2009. Revealing the physics of movement: Comparing the similarity of movement characteristics of different types of moving objects. *Computers, Environment and Urban Systems* 33(6): 419–434.

Elango, V. V. and R. Guensler. 2010. An automated activity identification method for passively collected GPS data. Paper presented at *3rd Conference on Innovations in Travel Modeling*, Tempe, Arizona, USA.

Fellendorf, M. and P. Vortisch. 2010. Microscopic traffic flow simulator VISSIM. In *Fundamentals of Traffic Simulation*, edited by J. Barceló, 63–93. New York, NY: Springer.

Franklin, S. and A. Graesser. 1996. Is it an agent, or just a program? A taxonomy for autonomous agents. Paper presented at *International Workshop on Agent Theories, Architectures, and Languages*, 21–35. Springer, Berlin Heidelberg.

Fried, M., J. Havens, and M. Thall. 1977. *Travel behavior—A synthesized theory*. No. Project 8–14 Final Report.

Gerlough, D. and M. Huber. 1975. *Traffic Flow Theory: A Monograph (No. 165)*. Transportation Research Board. Washington, DC.

Gong, L., H. Sato, T. Yamamoto, T. Miwa, and T. Morikawa. 2015. Identification of activity stop locations in GPS trajectories by density-based clustering method combined with support vector machines. *Journal of Modern Transportation* 23(3): 202–213. doi: 10.1007/s40534-015-0079-x.

Gonzalez, M. C., C. A. Hidalgo, and A.-L. Barabasi. 2008. Understanding individual human mobility patterns. *Nature* 453(7196): 779–782.

Gonzalez, P., J. Weinstein, S. Barbeau, M. Labrador, P. Winters, N. Georggi, and R. Perez. 2010. Automating mode detection for travel behavior analysis by using global positioning systems-enabled mobile phones and neural networks. *IET Intelligent Transport Systems* 4(1): 37–49.

Gudmundsson, J., M. van Kreveld, and F. Staals. 2013. Algorithms for hotspot computation on trajectory data. In *Proceedings of the 21st ACM SIGSPATIAL International Conference on Advances in Geographic Information Systems*, 134–143, Orlando, Florida.

Hägerstrand, T. 1970. What about people in regional science? *Papers of the Regional Science Association* 24: 7–24.

Hemminki, S., P. Nurmi, and S. Tarkoma. 2013. Accelerometer-based transportation mode detection on smartphones. In *Proceedings of the 11th ACM Conference on Embedded Networked Sensor Systems*, 13, Roma, Italy.

Hensher, D. A., J. M. Rose, and W. H. Greene. 2005. *Applied Choice Analysis: A Primer*. Cambridge, UK: Cambridge University Press.

Horni, A., K. Nagel, and K. W. Axhausen. 2016. *The Multi-Agent Transport Simulation MATSim*. Ubiquity, London.

Huang, Z. and P. Williamson. 2001. *A Comparison of Synthetic Reconstruction and Combinatorial Optimisation Approaches to the Creation of Small-Area Microdata*. Working paper, Department of geography, University of Liverpool, Liverpool, United Kingdom.

Huynh, N., M. R. Namazi-Rad, P. Perez, and M. J. Berryman. 2013. Generating a synthetic population in support of agent-based modeling of transportation in sydney. Paper presented at *20th International Congress on Modeling and Simulation*. Adelaide, Australia.

Jain, S., N. Ronald, and S. Winter. 2015. Creating a synthetic population: A comparison of tools. Paper presented at *3rd Conference of Transportation Research Group (CTRG) of India*. Kolkata, India.

Jariyasunant, J., A. Carrel, V. Ekambaram, D. Gaker, T. Kote, R. Sengupta, and J. L. Walker. 2011. The quantified traveler: Using personal travel data to promote sustainable transport behavior, University of California Transportation Center, Berkeley. Retrieved from: http://escholarship.org/uc/item/678537sx.

Jones, P. M. 1979. New approaches to understanding travel behaviour: The human activity approach. In *Behavioral Travel Modeling*, edited by D. A. Hensher and P.R. Stopher, 55–80. London: Redwood Burn Ltd.

Kutter, E. 1984. Integrierte Berechnung Städtischen Personenverkehrs— Dokumentation der Entwicklung eines Verkehrsberechnungsmodells für die Verkehrsentwicklungsplanung Berlin (West). [Integrated calculation of urban passenger traffic—Documentation of the development of a traffic calculation model for traffic development planning of Berlin (West)].

Kutter, E. and H-J. Mikota. 1990. Weiterentwicklung des Personenverkehrsmodells Berlin auf der Basis der Verkehrsentstehungsmatrix 1986 (BVG). [Further development of the passenger transport model of Berlin based on the traffic generation matrix 1986].

Kutter, E., H-J. Mikota, J. Rümenapp, and I. Steinmeyer. 2002. Untersuchung auf der Basis der Haushaltsbefragung 1998 (Berlin und Umland) zur Aktualisierung des Modells "Pers Verk Berlin /RPlan", sowie speziell der Entwicklung der Verhaltensparameter '86 - '98 im Westteil Berlins, der Validierung bisheriger Hypothesen zum Verhalten im Ostteil, der Bestimmung von Verhaltensparametern für das Umland. Entwurf des Schlussberichts im Auftrag der Senatsverwaltung für Stadtentwicklung Berlin. Berlin/Hamburg. [Exploration of the household travel survey in 1998 (Berlin and the surrounding area) to update the model "Pers Verk Berlin/RPlan", as well as the development of the behavioral parameters '86–'98 in the Western part of Berlin, the validation of previous hypotheses on behavior in the Eastern part, and the determination of behavioral parameters for the surrounding area].

Lari, Z. A. and A. Golroo. 2015. Automated transportation mode detection using smart phone applications via machine learning: Case study mega city of Tehran. Paper presented at *Transportation Research Board 94th Annual Meeting*, Washington, DC.

Lee, J. H., A. Davis, E. McBride, and K. G. Goulias. 2017. Exploring social media data for travel demand analysis: A comparison of Twitter, household travel survey, and synthetic population data in California. Paper presented at *Transportation Research Board 96th Annual Meeting*, Washington, DC.

Levy, S. 1992. *Artificial Life: The Quest for a New Creation*. New York: Random House Inc.

Li, S., S. Dragicevic, F. A. Castro et al. 2016. Geospatial big data handling theory and methods: A review and research challenges. *ISPRS Journal of Photogrammetry and Remote Sensing* 115: 119–33.

Liao, L., D. Fox, and H. Kautz. 2005. Location-based activity recognition using relational Markov networks. In *Proceedings of the 19th International Joint Conference on Artificial Intelligence*, 773–778, Edinburgh, Scotland.

Liao, L., D. Fox, and H. Kautz. 2007. Extracting places and activities from GPS traces using hierarchical conditional random fields. *The International Journal Of Robotics Research* 26(1): 119–134. doi: 10.1177/0278364907073775.

Lim, P. P. and D. Gargett. 2013. Population synthesis for travel demand forecasting. Paper presented at *Australasian Transport Research Forum (ATRF), 36th*, Brisbane, Queensland, Australia.

Long, J. A. 2016. Kinematic interpolation of movement data. *International Journal of Geographical Information Science* 30(5): 854–868. doi: 10.1080/13658816.2015.1081909.

McNally, M. G. and C. R. Rindt. 2008. The activity-based approach. In *Handbook of Transport Modeling*, edited by D. A. Hensher and K. J. Button, 55–73. Amsterdam, The Netherlands: Elsevier.

Miller, H. J. and S.-L. Shaw. 2001. *Geographic Information Systems for Transportation: Principles and Applications*. Oxford: Oxford University Press.

Miller, H. J. and S.-L. Shaw. 2015. Geographic information systems for transportation in the 21st Century. *Geography Compass* 9(4): 180–189.

Murakami, E. and D. P. Wagner. 1999. Can using global positioning system (GPS) improve trip reporting? *Transportation Research Part C: Emerging Technologies* 7: 149–165.

Ortúzar, J. de D., J. Armoogum, J.-L. Madre, and F. Potier. 2011. Continuous mobility surveys: The state of practice. *Transport Reviews* 31(3): 293–312.

Padgham, L., K. Nagel, D. Singh, and Q. Chen. 2014. Integrating BDI agents into a MATSim simulation. *Frontiers in Artificial Intelligence and Applications* 263: 681–686.

Palma, A. T., V. Bogorny, B. Kuijpers, and L. O. Alvares. 2008. A clustering-based approach for discovering interesting places in trajectories. In *Proceedings of the 2008 ACM Symposium on Applied Computing*, 863–868, Fortaleza, Ceara, Brazil.

Persula, M. 1999. Simulation of traffic systems—An overview. *Journal of Geographic Information and Decision Analysis* 3(1): 1–8.

Pinjari, A. R. and C. R. Bhat. 2011. Activity-based travel demand analysis. In *A Handbook of Transport Economics*, edited by A. D. Palma, 213–248. Cheltenham, England: Edward Elgar Pub.

Pinjari, A. R., N. Eluru, R. Copperman et al. 2006. Activity-based travel-demand analysis for metropolitan areas in Texas: CEMDAP models, framework, software architecture and application results. Texas Department of Transportation No. FHWA/TX-07/0-4080-8.

Raveau, S., A. Ghorpade, F. Zhao, M. Abou-Zeid, C. Zegras, and M. Ben-Akiva. 2016. Smartphone-based survey for real-time and retrospective happiness related to travel and activities. Paper presented at *Transportation Research Board 95th Annual Meeting*. Washington, DC.

Reddy, S., M. Mun, J. Burke, D. Estrin, M. Hansen, and M. Srivastava. 2010. Using mobile phones to determine transportation modes. *ACM Transactions on Sensor Networks* 6(2): 1–27. doi: 10.1145/1689239.1689243.

Rieser, M., K. Nagel, U. Beuck, M. Balmer, and J. Rümenapp. 2007. Truly agent-oriented coupling of an activity-based demand generation with a multi-agent traffic simulation. Paper presented at *Transportation Research Board 86th Annual Meeting*. Washington, DC.

Rocha, J. A. M. R., V. C. Times, G. Oliveira, L. O. Alvares, and V. Bogorny. 2010. DB-SMoT: A direction-based spatio-temporal clustering method. 2010. Paper presented at *5th IEEE International Conference Intelligent Systems*, London, UK.

Roorda, M. J., A. Shalaby, and S. Saneinejad. 2011. Comprehensive transportation data collection: Case study in the Greater Golden Horseshoe, Canada. *Journal of Urban Planning and Development* 137(2): 193.

Ryan, J., H. Maoh, and P. Kanaroglou. 2009. Population synthesis: Comparing the major techniques using a small, complete population of firms. *Geographical Analysis* 41(2): 181–203.

Safi, H., M. Mesbah, and L. Ferreira. 2013. ATLAS Project: Developing a mobile-based travel survey. In *Proceedings of the Australian Transportation Research Forum*, 2–4, Brisbane, Australia.

Santi, P., G. Resta, M. Szell, S. Sobolevsky, S. H. Strogatz, and C. Ratti. 2014. Quantifying the benefits of vehicle pooling with shareability networks. *Proceedings of the National Academy of Sciences* 111(37): 13290–13294.

Schuessler, N. and K. Axhausen. 2009. Processing GPS raw data without additional information. Paper presented at *Transportation Research Board 88th Annual Meeting*, Washington, DC.

Sevtsuk, A. and C. Ratti. 2010. Does urban mobility have a daily routine? Learning from the aggregate data of mobile networks. *Journal of Urban Technology* 17(1): 41–60.

Smith, L., R. Beckman, and K. Baggerly. 1995. *TRANSIMS: Transportation analysis and simulation system*. No. LA-UR--95-1641. Los Alamos National Lab., New Mexico.

Spaccapietra, S. and C. Parent. 2011. Adding meaning to your steps. In *Proceedings of the 30th International Conference on Conceptual Modeling*, 13–31, Brussels, Belgium.

Stenneth, L., K. Thompson, W. Stone, and J. Alowibdi. 2012. Automated transportation transfer detection using GPS enabled smartphones. In *2012 15th International IEEE Conference on Intelligent Transportation Systems Intelligent Transportation Systems (ITSC)*, 802–807. Anchorage, AK.

Stenneth, L., O. Wolfson, P. S. Yu, and B. Xu. 2011. Transportation mode detection using mobile phones and GIS information. In *Proceedings of the 19th ACM SIGSPATIAL International Conference on Advances in Geographic Information Systems*, 54–63. Chicago, IL.

Stopher, P. R. 2004. GPS, location, and household travel. In *Handbook of Transport Geography and Spatial Systems*, edited by D. A. Hensher, 432–449. Amsterdam, The Netherlands: Elsevier Ltd.

Stopher, P. R. 2008. Survey and sampling strategies. In *Handbook of Transport Modeling*, edited by D. A. Hensher and K. J. Button, 279–302. Amsterdam, The Netherlands: Elsevier.

Stopher, P. R., C. FitzGerald, and M. Xu. 2007. Assessing the accuracy of the Sydney household travel survey with GPS. *Transportation* 34(6): 723–741.

Stopher, P. R., E. Clifford, J. Zhang, and C. FitzGerald. 2008. Deducing mode and purpose from GPS data. Working paper, ITLS-WP-08-06, Australia: Institute of Transports and Logistics Studies, The University of Sydney.

Stopher, P. R., P. Bullock, and Q. Jiang. 2002. GPS, GIS and personal travel surveys: An exercise in visualization. Paper presented at *25th Australian Transport Research Forum*, Canberra, Australia.

Stopher. P. R., R. Alsnih, and C. G. Stecher et al. 2006. *Standardization of Personal Travel Surveys*. Washington DC: Transportation Research Board.

Torrens, P.M. 2004. Geosimulation, automata, and traffic. In *Handbook of Transport Geography and Spatial Systems*, edited by D.A. Hensher, 549–564. Amsterdam, London: Bingley, UK: Emerald, c2008.

Williamson, P. 2002. The census data system. In *Synthetic Microdata*, edited by P. Rees, D. Martin, and P. Williamson, 231–241. Chichester: Wiley.

Williamson, P., M. Birkin, and P. H. Rees. 1998. The estimation of population micro-data by using data from small area statistics and samples of anonymised records. *Environment and Planning A* 30(5): 785–816.

Wilson, A. G. and C. E. Pownall. 1976. A new representation of the urban system for modeling and for the study of micro-level interdependence. *Area* 8(4): 246–254.

Wise, S., A. Crooks, and M. Batty. 2016. Transportation in agent-based urban modeling. In *Agent Based Modeling of Urban Systems: First International Workshop, ABMUS 2016, Held in Conjunction with AAMAS, Singapore, Singapore, May 10, 2016, Revised, Selected, and Invited Papers*, edited by M. R. Namazi-Rad, L. Padgham, P. Perez, K. Nagel, A. Bazzan, 129–148. Cham: Springer.

Wolf, J., M. Oliveira, and M. Thompson. 2003. Impact of underreporting on mileage and travel time estimates: Results from global positioning system-enhanced household travel survey. *Transportation Research Record: Journal of the Transportation Research Board* 1854: 189–98.

Wolf, J., S. Schönfelder, U. Samaga, M. Oliveira, and K. W. Axhausen. 2004. Eighty weeks of GPS traces: Approaches to enriching trip information. *Transportation Research Board 83rd Annual Meeting*, Washington DC.

Wu, L., B. Yang, and P. Jing. 2016. Travel mode detection based on GPS raw data collected by smartphones: A systematic review of the existing methodologies. *Information* 7(4): 67.

Xiao, G., Z. Juan, and J. Gao. 2015. Travel mode detection based on neural networks and particle swarm optimization. *Information* 6(3): 522–535.

Xiao, Z., Y. Wang, K. Fu, and F. Wu. 2017. Identifying different transportation modes from trajectory data using Tree-based ensemble classifiers. *ISPRS International Journal of Geo-Information* 6(2): 57.

Xu, C., M. Ji, W. Chen, and Z. Zhang. 2010. Identifying travel mode from GPS trajectories through fuzzy pattern recognition. Paper presented at *International Conference on Fuzzy Systems and Knowledge Discovery (FSKD)*, Yantai, China.

Yalowitz, S. S. and K. Bronnenkant. 2009. Timing and tracking: Unlocking visitor behavior. *Visitor Studies* 12(1): 47–64.

Ye, X., K. Konduri, R. M. Pendyala, and B. Sana. 2009. A Methodology to match distributions of both household and person attributes in the generation of synthetic populations. Paper presented at *Transportation Research Board 88th Annual Meeting*, Washington, DC.

Zheng, Y., L. Liu, L. Wang, and X. Xie. 2008. Learning transportation mode from raw GPS data for geographic applications on the Web. Paper presented at *17th World Wide Web Conference*, Beijing, China.

Zheng, Y. T., Z.-J. Zha, and T.-S. Chua. 2012. Mining travel patterns from geotagged photos. *ACM Transactions on Intelligent Systems and Technology (TIST)* 3(3): 56.

Zhong, C., M. Batty, E. Manley, J. Wang, Z. Wang, and F. Chen. 2016. Variability in regularity: Mining temporal mobility patterns in London, Singapore and Beijing using smart-card data. *PloSOne* 11(2): e0149222.

Zimmermann, M., T. Kirste, and M. Spiliopoulou. 2009. Finding stops in error-prone trajectories of moving objects with time-based clustering. In *Intelligent Interactive Assistance and Mobile Multimedia Computing: International Conference, IMC 2009, Rostock-Warnemünde, Germany, November 9–11, 2009. Proceedings*, edited by D. Tavangarian, T. Kirste, D. Timmermann, U. Lucke and D. Versick, 275–286. Berlin, Heidelberg: Springer.

Zmud, J. 2003. Designing instruments to improve response. In *Transport Survey Quality and Innovation*, edited by P. Jones and P.R. Stopher, 89–108. Bingley, UK: Emerald.

9

Geography of Social Media in Public Response to Policy-Based Topics

Xinyue Ye, Shengwen Li, Adiyana Sharag-Eldin, Ming-Hsiang Tsou, and Brian Spitzberg

CONTENTS

9.1 Introduction

In today's increasingly connected world of virtual, perceived, and real spaces, social media data have been used to enhance decision making, understand customer behavior, improve operational efficiency and identify new markets (Leavey, 2013; Ye and He, 2016). With the development of Web 2.0, interactions through the Internet can provide an insight for decision makers to formulate better policy targeting the right groups of people, which has been a research goal pursued in many disciplines (Wu et al., 2016). The complexities of such interaction systems at various spatial, temporal, and semantic scales have posed both challenges and opportunities to researchers (Ye Huang, and Li, 2016). Policy makers aim to utilize social media data to comprehend the complex social problems at a variety of scales and to compete for attention and influence within it (Margetts, 2009; Leavey, 2013). In addition, social media data can also help improve the quality and timeliness of the evidence base that informs new public policy. More importantly, it is critical for policy makers to know which groups of people they are listening to, where they are from, and what those people's socioeconomic status is (Wang et al., 2016). In this case, geography, location information and associated socioeconomic status are critical data sources in addition to the text of social media data (Ye et al., 2016).

In the networked society, promoting sustainability and environmental justice requires the use of advanced technologies to understand complex challenges including urbanization and climate change (Alexander, 2014; Wang et al., 2015; Chong et al., 2016; Wang et al., 2016). Russell-Verma et al. (2016) conducted qualitative content analysis of online news articles and their associated readers' comments to examine public preferences for drought mitigation options in the south-east of England and their reasons. Kohl and Knox (2016) investigated multiple ways to understand how scientific operationalizations of drought interact with the politics of water management in Georgia in the southeastern United States, based on archival research, direct observation, and semistructured interviews with various interest groups. The Internet access and widely adopted social media platforms support the general public to express their opinions. Kohl and Knox (2016) argued that scientists and policymakers need to understand how knowledge of drought is developed through interactions between science, nature, and society. The policy issue will serve as an exemplary canvas upon which public attitudes will be written in online and social media messaging and other traditional news media. Social media influence personal opinion, which in turn influences broader public perceptions of social issues Ceron and Negri, 2016. In developing an opinion, many people do not use rational knowledge to make up their own opinions. People express the opinions of friends or those of opinion leaders that consist of prominent people, or politicians. According to Noelle-Neumann (1974), people learn about public opinion from media coverage especially when the news contain a contradiction. Hence, rigorous analysis of such data is likely to open up a rich context for advancing public policy interventions.

This research uses tweets to illustrate how social media memes reflect public response to social and policy-based topics, using a water bond proposition in California as the case exemplar. The increasing complexity of environmental issues demands place-based interdisciplinary thinking and approaches that make recommendations among diverse stakeholders at the policy, community, and organizational scales. Most social media related studies on policy issues in Europe and North America are on floods and forest fires (Wang et al., 2016), and there is a lack of research on drought (Klonner et al., 2016). We also aim to explore how the social media users are integrated into various local and topical contexts. The serious statewide drought in California has impacted both agriculture and the environment (Macon et al. 2016). California Proposition 1, the Water Bond (Assembly Bill 1471) as a legislatively-referred bond act, was voted and approved on the November 4, 2014 ballot in California. The previous measure known as Proposition 43 (Myers, 2014) was replaced by the new measure, which aims to improve the Water Quality, Supply, and Infrastructure. Proposition 1 tries to release $7.12 billion obligation bonds to sponsor state water supply infrastructure projects, appropriate money from the General Fund to pay off bonds, and require certain projects to provide matching funds from non-state sources in order to receive bond funds. California Proposition 1 was approved with a total of 4,771,350 people voting yes. This was 67.13% of the total population. We select this phenomon as the background of social measure because it is a social topic that generated certain controversy and debates examining the interplay between science and politics in the spatial context. The remainder of this chapter is structured as follows. The data section describes the social media data used in this study. The data Analysis section presents the analytical results. We then conclude our discussion with a summary of the findings and directions for further works.

9.2 Data

Using the Twitter Search Application Programming Interface (API) and based on the keywords of "water bond" and "vote," two groups of tweets were gathered respectively in California in 2014. These two words appeared frequently as hashtags and retweets. Another advantage of using Twitter is the ability of the program to track phrases, words in hashtags that are most mentioned and posted under the title of "trending topics." The hashtag convention allows users to search for the word contained in tweets that feature a specific character such as "#" for hashtags. The hashtags water bond and vote (#water bond and #vote) used in this project tracked Twitter messages using the same hashtag. The first group was from October 10 to November 20, and the second group was from November 7 to November 30. In the "water bond" dataset, 2.28% of tweets had geographical coordinates. At the same time, 2.33% of "vote" related tweets were geocoded. A meme is defined here as a unit of information that is spread (i.e., reproduced with some degree of fidelity, or diffused) from person-to-person through a social network. Once a meme is identified, it can be used to classify different types of social phenomena. Most memes go only one or two steps or nodes in a social network, but memes that are spread widely can be said to be viral in their diffusion pattern. Virality is a communication phenomenon in which thousands to millions share a meme over a relatively brief interval of time.

After removing duplicated tweets, there are 1530 unique "water bond" tweets, while there are 150,000 "vote" tweets. Seven columns are identified in each tweet: UID (unique identifier of the tweet), UserID (ID of the user who posted the tweet), City (the city name in the user's profile), Text (the content of the tweet with a maximum of 140 characters), Geo (the coordinates of the location where the tweet was posted), Location (place name such as city and state where the tweet was posted), Time (the day and time when the tweet was posted), and Localtime (the users' local time when the tweet was posted). Tweets contain a wealth of spatial information, which can help better understand the phenomena associated with the geographic location (Wang, Ye, and

Tsou, 2016). A meme is an analog of genes in biology; that is, a meme is any replicable message that spreads information or culture from person to person. When a word becomes viral and is retweeted to other users, a meme is developed. The definition of a meme is a unit of information that spread from person to person through the social network. Once a meme is identified, we can use it to classify different types of social phenomena (Ferrara et al., 2013). It could help us understand what people are concerned about over time.

With the proliferation of social media, messages generated and diffused from these outlets have become an important component of our daily lives with great potential for effectively distributing political messages, hazard alerts, or messages of other social functions (Wang et al., 2015). The users' participation ranges from political groups and individuals with strong political agendas, to common citizens with marginally-involved viewpoints. Individuals may be influenced by direct exposure of drought risk or spread the news to each other through the social media communication (Watts and Dodds, 2007). An advantage of using Twitter data is the ability of tweets to provide georeferenced and time-stamped information that support the application of geographic information systems (GIS) analytics as planning and decision-making tools (Andrienko et al., 2013; Wang et al., 2015).

9.3 Data Analysis

Tweets have clear spatial distribution patterns. As shown in Table 9.1, the relative attention on "water bond" was greater than that on "vote" tweets in some cities. Among those cities, Sacramento and Stockton are at the center of the drought area in California as shown in Figure 9.1. The spatial analysis of social media streams can help us understand the spatial distribution of public response. Such comparison reveals that the place-based situational awareness matters in the social media platform. Local residents are concerned about the immediate challenges in the vicinity. The drought highlights the city's vulnerability to this natural hazard and social media messages called for the attention of water shortages and the need for drought mitigation measures. Hence, virtual space expressed in Table 9.1 is highly related to the situation in the physical space shown in Figure 9.1. This can be very helpful for the policy makers with a full awareness of public concern.

From another perspective, we can map the spatial distribution of "water bond" tweets at a finer scale using coordinates to compare the distribution of tweets across space. In order to reduce the impact of population, the number of tweets per one thousand persons were calculated with a one-mile resolution and a five-mile search radius (Figure 9.1).

Each tweet with location information can be regarded as a geographical entity. Figure 9.1 reveals a large difference across cities, with the highest value

TABLE 9.1

Difference in Distribution of Tweets in Cities

City	"Water Bond" (%)	"Vote" (%)
Sacramento	15.49	1.23
Stockton	3.27	0.31
Indio	2.16	0.32
Tulare	0.65	0.10
Riverside–San Bernardino	2.88	0.80
Fresno	3.66	1.14
Fairfield	0.92	0.31
San Francisco–Oakland	22.94	8.60
Salinas	0.65	0.27
Bakersfield	0.65	0.35
Santa_Rosa	0.26	0.14
San Diego–Chula Vista	10.52	6.67
San Jose–Freemont	4.84	3.09
Carlsbad–Oceanside–Escondido	0.20	0.42
Los Angeles–Long Beach	29.48	68.52
Oxnard	0.13	0.32
Santa Ana–Anaheim–Irvine	1.31	6.89
El_Centro	0.00	0.37
Lancaster	0.00	0.15

of 12.41 tweets per one thousand persons in Los Angeles-Long Beach. This result demonstrates that the topic of "water bond" drew more attention in coastal cities. As Klonner et al. (2016) argued, "integrating local knowledge, personal experience and up-to-date geoinformation indicates a promising approach for the theoretical framework and the methods of natural hazard analysis." California's coastal cities host large amounts of residents, where water shortage and environmental injustice are larger concerns for sustainable urban growth. November 4th 2014, the Water Bond Referendum day, was a key date of tweeting behavior. The number of "water bond" tweets reached their peak on this day. Distribution of tweets over time can be further analyzed using the granularity of hours. Table 9.1 and Figure 9.1 reflect Adams's (2011) statement that the social space of communication as places in media is an abstract, theoretical, and production-oriented spaces involving the formal plans and abstract blueprints of powerful actors whose formalizations of space control actions. "Place" captures the idea of deeply layered subjective experience grounded in the particularity of local conditions and discourses, whereas "space" implies potential as well as actual movement of bodies, goods, capital, information, and communication (Edwards and Usher, 2007).

Analyzing the trends of tweets can detect the changes in people's concerns during different phases of events and attitudes towards different types of

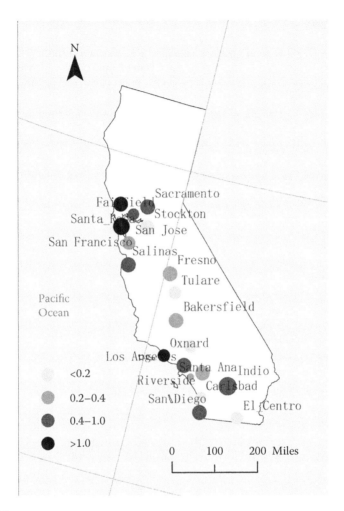

FIGURE 9.1
Spatial distribution of tweets per one thousand persons.

events (Wang et al., 2015). Because tweets relevant to certain topics tend to fluctuate over time, the overall trend of concerns towards certain topics can be assessed. Figure 9.2 reveals a huge difference between the specific topic of "water bond" and the general topic of "vote" at the temporal resolution of hours. Water bond attracted lots of discussion at late night, while voting issues consumed most of dinner time. Most retweeting activities occurred at late night for "water bond," while lunch and dinner times witnessed many more retweets for "vote."

Tweets have been posted across various platforms across many user affordances. The digital technology introduces a new momentum for the social media users to make a connection with other users. The users will select from varieties of the networking program applications that are appropriate for their

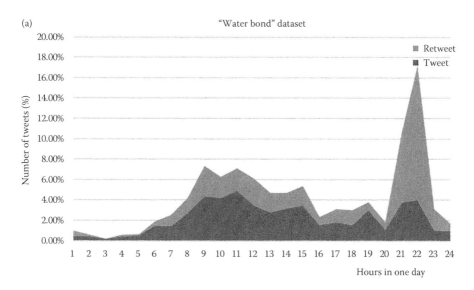

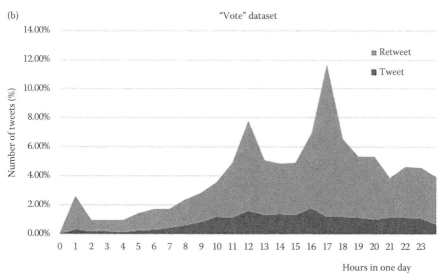

FIGURE 9.2
Distribution of tweets over hours (a) water bond; (b) vote.

activities (Kaplan and Haenlein, 2010). Tables 9.2 and 9.3 reveal that people tend to use computers to post more specific topics such as "water bond," compared to the preferences for mobile phone in tweeting general posts related to "vote." It is interesting that "water bond" is posted more through web clients, partially due to this specific topic needing more web search in the computer. However, mobile devices overwhelm other selections in sending "vote" related tweets, because this topic is more general and less specific.

TABLE 9.2

Top 10 Platforms in "Water Bond" Dataset

Source	Number of Tweets	Percent
Twitter Web Client	374	24.44
Twitter for iPhone	287	18.76
Hootsuite	118	7.71
TweetDeck	118	7.71
Twitter for Android	90	5.88
Twitter for Websites	78	5.10
dlvr.it	75	4.90
Twitter for iPad	74	4.84
Twitterfeed	53	3.46
RoundTeam	33	2.16

TABLE 9.3

Top 10 Platforms in "Vote" Dataset

Source	Number of Tweets	Percent
Twitter for iPhone	58,822	39.21
Twitter for Android	31,412	20.94
Twitter Web Client	24,856	16.57
Twitter for iPad	6878	4.59
TweetDeck	6693	4.46
Twitter for Websites	3492	2.33
Twitter for Android Tablets	2389	1.59
Hootsuite	1355	0.90
Twitter for Windows Phone	1342	0.89
Facebook	1186	0.79

9.4 Conclusions

As a computer-mediated tool, social media platform enables people to create and exchange information, ideas, pictures, and videos in virtual communities and networks (Li et al., 2017). As one of the world's major social media platforms, Twitter allows individuals to timely access to diffuse messages from a particular location at a given moment for a specific policy issue, along with a growing pronounced concern for the relative social justice of choices. Users can broadcast brief text updates to public or to a selected group of contacts facilitating the actions and interactions among individuals and/or agencies (Lai, She, and Ye, 2015; Yang, Ye, and Sui, 2016). For a social debate, traditional intellectual inquiry and survey take a long time to complete with only a limited number of participants. Social media are replacing face-to-face social interaction and redefining the diffusion of information (Ratkiewicz et al., 2011; Chen et al., 2016). Tweets related to water bond in California have been explored from spatial, temporal, and topical perspectives in this chapter, in order to demonstrate how people connect, network, and promote opinions. An analytical framework is developed to derive effective, validated, and usable information.

Public policy decision making is undergoing rapid transition towards analyzing ever-increasing amounts of large-scale geosocial response data collected from citizens and stakeholders. Investigating such response dynamics needs an analytical framework within which effective solutions might be developed in an interdisciplinary, collaborative, and timely manner (Ye, Huang, and Li, 2016). Compared to the general vote related tweets, geo-tagged "water bond" tweets were more attached to where the drought area was located. Residents in coastal area are more willing to express their thought in a brief message related to "water bond" in public. November 4th 2014, the Water Bond Referendum day, served as the turning point of the tweeting behavior dynamics. There were far fewer tweets before and after this peak day. People were also more likely to assign different time slots during the day to discuss specific or general policy issues. This study reveals that users tent to spend more time at late night to post tweets on more specific topics that need more work in web searching (Tsou, 2013). The temporal and textual phrase propagated on the tweets among the individuals with similar propensities occurs through the continuous interaction of messages on a daily basis (Ye and Lee, 2016). In addition, the preference of various platforms matter in different topics, presumably because users might need references and support materials for certain topics. It will be a promising path to integrate the measurement of space, time and topic, in order to better understand how perceived space and real space affect each other (Crampton, 2013; Tsou, 2015; Shaw et al., 2016). Even in instances where the knowledge on drought is different, stakeholders can still change the societal context, as the green industry did in Georgia in 2009.

This chapter contributes to the literatures as an empirical public policy related study to utilize social media and volunteered geographic data to explore human behavior and social phenomena. This study generates important lessons about the structured use of social media sources for understanding geography of social media in public response to social topics. The next step will be identifying opinion leaders in this water bond tweets and examine how message diffusion occurs across space and over time, based on ever-increasing amounts of large-scale geosocial data and computing power. As Lefebvre (1991) argued, the representation of space or the social space of communication is occupied by artists, writers, and philosophers. In the networked world today, the space of communication is also shared by the Internet bloggers, celebrities, and political figures. These opinion leaders will attract more intense commentary and divide opinion to a greater extent than common users.

The second stage of analysis can involve an online survey and face-to-face interview with the residents of the impacted communities. At the same time, it is important to differentiate the supply side interventions from demand side strategies in understanding the dialogue around mitigation options. This can be used to explore how stakeholders might leverage the complexity of drought to achieve their political goals and adjust water management policy during times of drought. Mixed approaches involving rich and complex spatial, temporal, network, and semantics data are needed to facilitate transformative geospatial discovery for enabling effective and timely solutions to the challenging environmental problems. Advanced natural language processing (NLP) tools are needed to accompany the pre-selection of keywords in tweets. NLP will help the data screening by identifying which concepts a word or phrase stands for and knowing how to link those concepts together in a meaningful way. There are nevertheless important limitations to reliance on social media as a proxy for public opinion. In this study, there were not enough data to investigate the roles of geotechnical external factors such as technological penetration of the population to localizable clusters of tweets. Geotechnical factors in this case study might have included population demographics affecting the adoption of Twitter (which is a "younger generation" medium), the concentration of major media outlets, urban versus rural divisions that index political ideologies, and perhaps even neighboring state politics or shared urban metropolitan areas (Spitzberg, 2014).

Funding

The author(s) disclosed receipt of the following financial support for the research, authorship, and/or publication of this article: This material is based upon work supported by the National Science Foundation under Grant No. 1416509, project titled "Spatiotemporal Modeling of Human Dynamics Across Social Media and Social Networks." Any opinions, findings, and conclusions or recommendations expressed in this material are those of the authors and do not necessarily reflect the views of the National Science Foundation.

References

Adams, P. C. 2011. A taxonomy for communication geography. *Progress in Human Geography* 35 (1): 37–57.

Alexander, D. E. 2014. Social media in disaster risk reduction and crisis management. *Science and Engineering Ethics* 20 (3): 717–733.

Andrienko, G., N. Andrienko, H. Bosch, T. Ertl, G. Fuchs, P. Jankowski and D. Thom. 2013. Thematic patterns in georeferenced tweets through space-time visual analytics. *Computing in Science and Engineering* 15 (3): 72–82.

Ceron, A. and F. Negri. 2016. The "social side" of public policy: Monitoring online public opinion and its mobilization during the policy cycle. *Policy and Internet* 8 (2): 131–147. doi:10.1002/poi3.117.

Chen, X., G. Elmes, X. Ye and J. Chang. 2016. Implementing a real-time Twitter-based system for resource dispatch in disaster management. *GeoJournal* 81 (6): 863–873.

Chong, Z., C. Qin and X. Ye. 2016. Environmental Regulation, Economic Network and Sustainable Growth of Urban Agglomerations in China. *Sustainability* 8 (5): 1–21.

Crampton, J. W. 2013. Beyond the geotag: situating 'big data' and leveraging the potential of the geoweb. *Cartography and Geographic Information Science* 40 (2): 130–139.

Edwards, R. and R. Usher, 2007. *Globalisation & Pedagogy: Space, Place and Identity.* London: Routledge.

Ferrara, E., M. JafariAsbagh, O. Varol, V. Qazvinian, F. Menczer and A. Flammini, 2013, August. Clustering memes in social media. In *Advances in Social Networks Analysis and Mining (ASONAM), 2013 IEEE/ACM International Conference on*, 548–555. IEEE.

Kaplan, A. M. and M. Haenlein, 2010. Users of the world, unite! The challenges and opportunities of Social Media. *Business Horizons* 53 (1): 59–68.

Klonner, C., S. Marx, T. Usón, J. Porto de Albuquerque and B. Höfle. 2016. Volunteered geographic information in natural hazard analysis: a systematic literature review of current approaches with a focus on preparedness and mitigation. *ISPRS International Journal of Geo-Information* 5 (7): 103.

Kohl, E. and J. A. Knox, 2016. My Drought is Different from Your Drought: A Case Study of the Policy Implications of Multiple Ways of Knowing Drought. *Weather, Climate, and Society* 8 (4): 373–388.

Lai, C., B. She and X. Ye. 2015. Unpacking the Network Processes and Outcomes of Online and Offline Humanitarian Collaboration. *Communication Research.* doi:10.1177/0093650215616862

Leavey, J. 2013. *Social Media and Public Policy: What is the Evidence.* Publication: Alliance for Useful Evidence.

Lefebvre, H. 1991. *The Production of Space.* Trans Donald, Nicholson-Smith. Oxford: Blackwell, 26–38.

Li, Q., W. Wei, N. Xiang, D. Feng, X. Ye and Y. Jiang. 2017. Social Media Research, Human Behavior, and Sustainable Society. *Sustainability* 9 (3): 384.

Macon, D. K., S. Barry, T. Becchetti, J. S. Davy, M. P. Doran, J. A. Finzel, … D. E. Lancaster. 2016. Coping With Drought on California Rangelands. *Rangelands* 38 (4): 222–228.

Margetts, H. Z. 2009. The Internet and public policy. *Policy & Internet* 1 (1): 1–21.

Myers, J. 2014. New Ballot Numbers For November's Water. *Budget Propositions.* Retrieved January 1, 2017, from https://ww2.kqed.org/news/2014/08/12/new-ballot-numbers-for-novembers-water-budget-propositions/.

Noelle-Neumann, E. 1974. The spiral of silence a theory of public opinion. *Journal of Communication* 24 (2): 43–51.

Ratkiewicz, J., M. Conover, M., Meiss, B., Gonçalves, S., Patil, A., Flammini, and F. Menczer. 2011, March. Truthy: mapping the spread of astroturf in microblog streams. In *Proceedings of the 20th International Conference Companion on World Wide Web*, 249–252). ACM.

Russell-Verma, S., H. M. Smith and P. Jeffrey, 2016. Public views on drought mitigation: Evidence from the comments sections of on-line news sources. *Urban Water Journal* 13 (5): 454–462.

Shaw, S., M. Tsou and X. Ye. 2016. Human Dynamics in the Mobile and Big Data Era. *International Journal of Geographical Information Science* 30 (9): 1687–1693.

Spitzberg, B. H. 2014. Toward a model of meme diffusion (M3D). *Communication Theory* 24 (3): 311–339.

Tsou, M.-H. 2013. Mapping social activities and concepts with social media (Twitter) and web search engines (Yahoo and Bing): a case study in 2012 US Presidential Election. *Cartography and Geographic Information Science* 40 (4): 337–348.

Tsou, M.-H. 2015. Research challenges and opportunities in mapping social media and Big Data. *Cartography and Geographic Information Science* 42 (Suppl. 1): 70–74.

Wang, Y., W. Jiang, S. Liu, X Ye and T. Wang. 2016. Evaluating trade areas using social media data with a calibrated Huff model. *ISPRS International Journal of Geo-Information* 5 (7): 112.

Wang, Y., T., Wang, X. Ye, J. Zhu and J. Lee. 2015. Using Social Media for Emergency Response and Urban Sustainability: A Case Study of the 2012 Beijing Rainstorm. *Sustainability* 8 (1): 25.

Wang, Z., X Ye and M. Tsou. 2016. Spatial, temporal, and content analysis of Twitter for wildfire hazards. *Natural Hazards.* doi:10.1007/s11069-016-2329-6

Watts, D. J. and P. S. Dodds, 2007. Influentials, networks, and public opinion formation. *Journal of Consumer Research* 34 (4): 441–458.

Wu, C., X. Ye, F. Ren, Y. Wan, P. Ning and Q. Du, 2016. Spatial and Social Media Data Analytics of Housing Prices in Shenzhen, China. *PLOS One* 11 (10): 1–19.

Yang, X., X. Ye and D. Z. Sui. 2016. We Know Where You Are: In Space and Place-Enriching the Geographical Context through Social Media. *International Journal of Applied Geospatial Research* 7 (2): 61–75.

Ye, X. and C. He. 2016. The New Data Landscape for Regional and Urban Analysis. *GeoJournal* 81 (6): 811–815.

Ye, X. and J. Lee. 2016. Integrating geographic activity space and social network space to promote healthy lifestyles. *ACMSIGSPATIAL Health GIS* 8 (1): 24–33.

Ye, X., Q. Huang and W. Li. 2016 Integrating Big Social Data, Computing, and Modeling for Spatial Social Science. *Cartography and Geographic Information Science* 43 (5): 377–378.

Ye, X, S. Li, X. Yang and C. Qin. 2016. Use of Social Media for Detection and Analysis of Infectious Disease in China. *ISPRS International Journal of Geo-Information.* doi:10.3390/ijgi5090156.

10

Geospatial Data Streams

Zdravko Galić

CONTENTS

10.1 Introduction

The recent rapid development of wireless communication, mobile computing, global navigation satellite systems (GNSS), and spatially enabled sensors enables the exponential growth of available spatiotemporal data produced continuously at high speed. Owing to these advancements, a new class of monitoring applications has come into focus, including real-time intelligent transportation systems, traffic monitoring, and mobile objects tracking. These new *information flow processing (IFP)* application domains need to process huge volumes of geospatial data arriving in the form of continuous data streams. IFP applications are pushing traditional database technologies beyond their limits due to massively increasing data volumes and demands for real-time processing. Data stream management systems (DSMSs) have been developed by the database community to query and summarize continuous data streams for further processing. Owing to its pure relational paradigms, DSMSs have rudimentary geospatial processing capabilities. Geospatial stream processing refers to a class of software systems for processing high-volume geospatial data streams with very low latency, that is, in near real time. DSMSs are oriented toward processing large data streams in near real time. Despite the differences between these two classes of management systems, DSMSs resemble DBMSs; they process data streams using SQL, SQL-like expressions, and operators defined by relational algebra.

Geospatial data streams, that is, real-time, transient, time-varying sequences of geospatial data items, generated by embedded positioning sensors demonstrate at least two Big Data core features, *volume* and *velocity*. Increasingly, a dominant approach is to leverage in-memory computing over a cluster of commodity hardware. Similar to centralized DSMSs, existing distributed in-memory query engines and their processing models are predominantly based on relational paradigms and continuous operator models without explicit support for geospatial queries. There is a clear need for a highly scalable data stream computing framework that can operate at high data rates and process massive amounts of large geospatial data streams.

The goal of this chapter is twofold. First, to give an insight into geospatial stream processing at the conceptual level, that is, exclusively from the user's perspective, using a declarative, SQL-based approach. Second, to present a novel, in-memory parallel, and distributed prototype that supports real-time processing and analysis of large geospatial data streams.

10.2 From Databases to Data Streams

Database management systems (DBMSs) have been researched and used for a wide range of applications for over three decades. They have been used as a simple but effective warehouse of business data in applications that require persistent data storage and complex querying. A database consists of a finite, persistent set of objects that are relatively static, with insertions, deletions, and updates occurring less frequently than queries. Queries expressed in a query language such as SQL are executed when posed and the answers reflect the current state of the database. Over the years, it has become obvious that many applications involving geospatial data need extended and specialized DBMS functionalities. Geospatial databases evolved from traditional DBMS and have been successfully implemented as an extension of DBMS based on object-oriented or object-relational paradigms. More recently, spatiotemporal databases and their specific subclass called moving object databases have been an active area of research with a few available research prototypes. DBMSs have proven to be well suited to the organization, storage, and retrieval of finite, persistent geospatial datasets.

The new IFP application domains have forced an evolution of data processing paradigms, moving from DBMSs to DSMSs.

Data streams and new stream-oriented monitoring applications have the following characteristics and processing requirements:

- A data stream is an ordered, potentially unbounded sequence of data items called *stream elements*.
- Stream elements are generated by an active external source.
- The ordering is *implicit* if defined by the arrival time at the system.
- The ordering is *explicit* if stream elements provide generation time, that is, timestamps indicating their generation by an active external source.
- Stream elements are pushed by an active external source and arrive continuously at the system.
- The DSMS has neither control over the arrival order or arrival rate of stream elements.
- Stream elements are accessed sequentially; therefore, a stream element that has already arrived and has already been processed cannot be retrieved without being explicitly stored.
- It is possible to combine real-time processing with historical, persistent data in DBMS.
- Stream-oriented SQL enables real-time and historical analysis in a single (SQL-like) paradigm.
- A query over data streams runs continuously and returns new results as a new stream element arrives.

Traditional DBMSs employ a store-and-then-query data processing paradigm, that is, data are stored in the database and ad hoc queries are answered in full, based on the current snapshot of the database (Figure 10.1).

They could be used for data stream processing by loading data streams into persistent relations and repeatedly executing the same ad hoc queries over them. This approach requires that data streams need to be persisted on secondary storage devices, that is, disks with high latency, before they can be accessed and processed by a DBMS in the main memory. The mismatch between high latency of secondary storage and low latency of main memory adds considerable delay in response time that is not acceptable to many monitoring applications. IFP applications do not readily fit the traditional DBMS model and its query-driven,* pull-based processing paradigm, in which dynamic, transient ad hoc queries are typically specified, optimized, and processed once over relatively static, persistent data.

---

* Information flow processing.

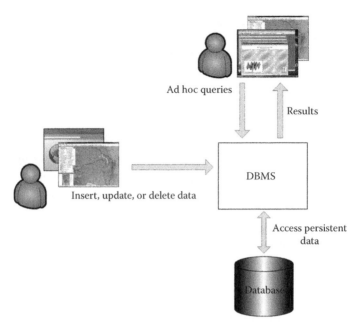

FIGURE 10.1
Query processing in DBMSs.

A data-driven, push-based processing paradigm in DSMSs is complementary to DBMSs: the same static, persistent queries are processed continuously over transient, dynamic, frequently changing data (Figure 10.2).

A data stream processing model views IFP as an evolution of data processing as supported by traditional DBMSs. Although DSMSs have their roots in DBMSs, they present significant differences: DBMSs are designed to work on persistent data where updates are relatively infrequent, while DSMSs are specialized in processing transient data that are continuously updated. While DBMSs run queries just once to return a complete answer, DSMSs execute the same standing queries, which run continuously and provide updated answers as new data arrive. Most DSMSs follow an integrated query processing approach that runs SQL continuously and incrementally over data before the data are stored in the database. Despite their differences, DSMSs resemble DBMSs, especially in the way they process incoming data through a sequence of transformations based on standard SQL operators; all the operators are defined by relational algebra. Table 10.1 gives an overview of differences between DSMSs and DBMSs.

It is worth noting that three DBMS variants are related to DSMSs:

- Real-time DBMSs provide all features of traditional DBMSs, while at the same time enforcing applications' real-time transaction constraints. Transaction processing in real-time DBMSs focuses on

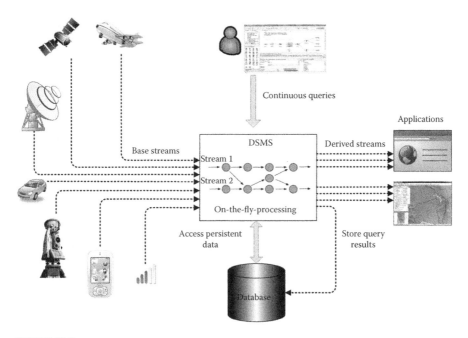

FIGURE 10.2
Query processing in DSMS.

enforcing time constraints of transactions and ensuring temporal consistency of data.

- In-memory DBMSs eliminate disk access by storing all data in memory and remove logical processes that are no longer necessary (i.e., caching), resulting in a relatively small code footprint.

- Embedded DBMS is an integral part of the application or application infrastructure, and runs with or as part of the application in embedded systems. Instead of providing full features of traditional DBMS, embedded DBMS provides minimal functionality such as indexing, concurrency control, logging, and transactional guarantees.

TABLE 10.1

Differences Between DSMSs and DBMSs

	DSMS	DBMS
Data	Transient streams	Persistent relations
Update rates	High	Low
Data access	Sequential, one-pass	Random
Queries	Continuous	*Ad hoc*, one-time
Query results	Exact or approximate	Exact
Latency	Low	High
Processing model	Data-driven (push-based)	Query driven (pull-based)

Although some of the goals for their development were similar to those of DSMSs, they are not able to meet the requirements of IFP applications. It is also important to note that DSMSs alone cannot entirely cover the needs of IFP: being an extension of DBMSs, they focus on producing query answers which are continuously updated to adapt to the constantly changing contents of their input data. This limitation originates from the nature of DSMSs: they are generic systems that leave the responsibility of associating semantics of the data being processed to their clients. Complex event processing (CEP) systems are capable of processing nongeneric data, which comprise event notifications coming from different external sources. CEP systems associate precise semantics with the notifications of events in the external world; the CEP engine is responsible for filtering and combining event notifications to understand what is occurring in terms of higher-level events. The detection and notification of complex patterns of events involving sequences and ordering relations are usually out of the scope of DSMSs. Therefore, CEP systems rely on the ability to specify composite events through event patterns that match incoming event notifications based on their content and ordering relationships.

10.3 Geospatial Continuous Queries

When processing geospatial data streams, there are two inherent temporal domains to consider:

- *Event time**—the time when the event itself occurred in the real world.
- *Processing time*—the current time according to the system clock when an event is observed during processing.

Most DSMSs are based on extensions of the relational model and corresponding query languages. Consequently, data stream items can be viewed as relational tuples with one important distinction: they are *time ordered*. The ordering is defined either explicitly by event time or implicitly by processing time. We define time domain, time instant, and time interval as follows.

Definition 10.1

(*Time Domain*) A time domain $\mathbb{T}$ is a pair $(T; \leq)$ where T is a non-empty set of discrete time instants and $\leq$ is the total order on T.

---

* The terms *event time* and *valid time* are often used interchangeably.

Definition 10.2

(*Time Instant*) A time instant τ is any value from T, i.e., $\tau \in T$.

Definition 10.3

(*Time Period*) A time period represents extended time defined by the temporal positions of the time instants at which it begins (τ_{begin}) and ends (τ_{end}).

Discrete time domain implies that every time instant has an immediate successor (except the last, if any) and immediate predecessor (except the first, if any).

Definition 10.4

(*Time Interval*) A time interval consists of all distinct time instants $\tau \in T$ and could be open, closed, left-closed, or right-closed:

$$(\tau_{begin}, \tau_{end}) = \{\tau \in T \mid \tau_{begin} < \tau < \tau_{end}\},$$

$$[\tau_{begin}, \tau_{end}] = \{\tau \in T \mid \tau_{begin} \leq \tau \leq \tau_{end}\},$$

$$[\tau_{begin}, \tau_{end}) = \{\tau \in T \mid \tau_{begin} \leq \tau < \tau_{end}\},$$

$$(\tau_{begin}, \tau_{end}] = \{\tau \in T \mid \tau_{begin} < \tau \leq \tau_{end}\}.$$

Geospatial data streams have two distinct features that differentiate them from conventional data streams based on relational models:

- The event time of a data stream tuple is defined by the temporal attribute A_θ.
- The shape and location of an object of interest described by a data stream tuple is defined by the spatial attribute A_σ.

The first feature implies that each data stream tuple has an event timestamp generated by the source, which classifies geospatial streams into a class of explicitly timestamped data streams.

The geospatial domain is a set of homogeneous object structures (values) that provides a fundamental abstraction for modeling the geometric structure of real-world phenomena in space. Points, lines, polygons, and surfaces are the most popular and fundamental abstractions for mapping a geometric structure from 3D space into 2D space. To locate an object in space, the embedding space must be defined. The formal treatment of spatial domain requires a definition of mathematical space. Although Euclidean $\mathbb{R}^2$ embedding space seems to be dominant, in some cases other spaces (metric, vector, and topological) are more important.

Simple object structures (point, continuous line, and simple polygon) are not closed under the geometric set operations (difference, intersection, and union). This means that geometric set operations can produce *complex* spatial objects (multi-points, multi-lines, multi-polygons, polygons with holes, etc.). For this reason, spatial domain should include spatial objects with complex structures.

Definition 10.5

(*Spatial Domain*) A spatial domain $\mathbb{D}_\sigma$ is a set of spatial objects with simple or complex structures.

Complex spatial objects may appear to be overwhelming in the DSMS context. However, we want to rely on and explore extensive research on abstract spatial data types in spatial DBMS and GIS. We are going to follow abstract spatial data type frameworks, where the internal structure of a spatial object is hidden from the user and can only be accessed through a set of predefined operations (Figure 10.3). A spatial domain, even though only consisting of simple object structures, is not atomic but rather structured, and consequently, geospatial data streams rely on object-relational (or object-oriented) paradigms.

Having defined the time and spatial domains, we define a geospatial data stream schema.

Definition 10.6

(*Geospatial Data Stream Schema*) A geospatial data stream schema Σ_γ is represented as a set of attributes $\langle A_1, A_2, \ldots, A_n \rangle$ of finite arity n. One of the attributes (denoted by A_σ) has an associated spatial domain $\mathbb{D}_\sigma$, and one of the attributes (denoted by A_θ) has an associated temporal domain $\mathbb{D}_\theta$, that is, $\mathbb{T}$. The values of other $n - 2$ attributes are drawn either from atomic type domain $\mathbb{D}_{\alpha_i}$ or complex type domain $\mathbb{D}_{\chi_i}$.

A geospatial data stream is defined in the following way.

Definition 10.7

(*Geospatial Data Stream*) A geospatial data stream S_γ is a possibly infinite sequence of tuples belonging to the schema of Σ_γ and ordered by the increasing values of A_θ.

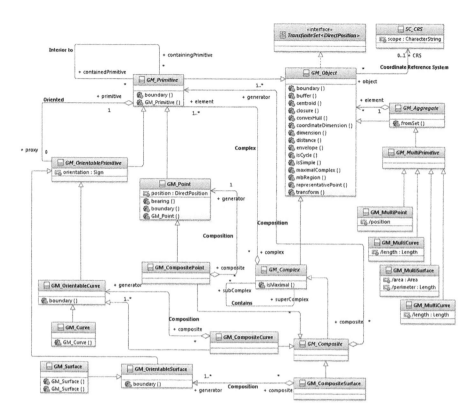

FIGURE 10.3
ISO 19107—Abstract spatial data types.

A geospatial stream tuple t_γ represents an *event*, that is, instantaneous fact capturing information that occurred in the real world at time instant τ, defined by an event timestamp. Event timestamps offer a unique time indication for each tuple and therefore cannot be either undefined (i.e., an event timestamp cannot have a *null* value) or mutable. In the following, we consider explicitly timestamped data streams, ordered by the increasing values of their event timestamps.

Definition 10.8

(*Temporal Order*) A temporal order is surjective (many-to-one) mapping $f_\Omega \colon \mathbb{D}_{S_\gamma} \to \mathbb{T}$ from data type domain D_{S_γ} of the tuples belonging to a data stream S_γ to time domain $\mathbb{T}$, such that the following holds:

1. Timestamp existence: $\forall s \in S_\gamma, \exists \tau \in \mathbb{T} \mid f_\Omega(s) = \tau$.
2. Timestamp monotonicity: $\forall s_1, s_2 \in S_\gamma, \textit{if } s_1.A_\Theta \leq s_2.A_\Theta \text{ then } f_\Omega(s_1) \leq f_\Omega(s_2)$.

A DSMS can tag each stream tuple with its arrival timestamp using the system's local clock. A stream with system timestamps can be processed like a regular stream with application event timestamps, but we should be aware that application event time and system time are not necessarily synchronized.

We distinguish between *raw* streams produced by the sources and *derived* streams produced by continuous queries and their operators. In either case, we model individual stream elements as object-relational tuples with a fixed geospatial schema.

The raw streams are an essential input for a broad range of applications such as traffic management and control, routing, and navigation. To become useful, the raw streams must be related to the underlying transportation network by means of map-matching algorithms. For example, map-matching is one of the key operations in Intelligent Transportation Systems; a map-matching user-defined function (UDF) could be applied on a raw stream to produce a derived stream.

10.3.1 Running Example

As a running example, let us consider the raw stream generated by GPS and speed sensors embedded into a mobile object:

```
CREATE STREAM gpsStream (
     id VARCHAR(8),
     lat REAL,            // Latitude
     lon REAL,            // Longitude
     elevation SMALLINT, // Ellipsoidal height
     speed REAL,                 // Speed [km/h]
     timestamp TIMESTAMP VALIDTIME
)
ORDERED BY timestamp;
ALTER STREAM gpsStream ADD WRAPPER gpsWrapper;
```

Streams might have multiple attributes of TIMESTAMP type, but only one should have a VALIDTIME constraint. This constraint implicitly determines the read-only attribute by which the stream is ordered. Wrappers are user-defined data acquisition functions that transform the sequence of bytes into a raw stream, and ALTER STREAM associates the raw stream with a wrapper.

An example of a sequence of gpsStream tuples (Figure 10.4):

```
...
"W-45084A" 48.20781333 16.43832500 221 73.2 2016-10-19 11:50:30
"W-45084A" 48.20795500 16.43853167 221 79.2 2016-10-19 11:50:31
"W-45084A" 48.20809167 16.43873667 220 77.4 2016-10-19 11:50:32
"W-45084A" 48.20823500 16.43894667 220 80.3 2016-10-19 11:50:33
"W-45084A" 48.20838167 16.43916333 220 82.5 2016-10-19 11:50:34
"W-45084A" 48.20851667 16.43936000 220 75.4 2016-10-19 11:50:35
```

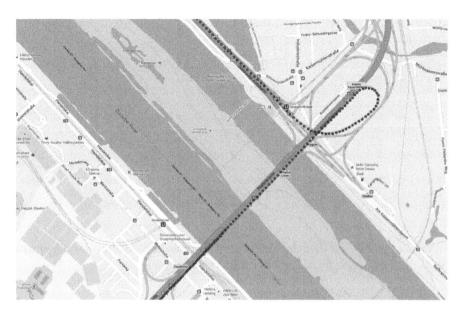

FIGURE 10.4
Visualization of a geospatial data stream.

```
"W-45084A" 48.20865833 16.43957167 220 80.1 2016-10-19 11:50:36
"W-45084A" 48.20881000 16.43978667 219 83.6 2016-10-19 11:50:37
"W-45084A" 48.20894667 16.43997667 219 74.7 2016-10-19 11:50:38
"W-45084A" 48.20908667 16.44017833 219 77.8 2016-10-19 11:50:39
"W-45084A" 48.20923167 16.44038333 218 79.8 2016-10-19 11:50:40
"W-45084A" 48.20937667 16.44059167 218 80.5 2016-10-19 11:50:41
...
```

A derived stream with a position modeled as a point on the WGS84 ellipsoid can be quite natural for applications involving geospatial objects whose movement is related to the Earth's surface (airplanes, tankers, combat aircraft, cruise missiles, and drones, among others):

```
CREATE STREAM mobileObjectWGS84 AS
SELECT id,
       SetSRID(Point(lon,lat,elevation),4326)::GEOGRAPHY
       AS wgsPosition,
       speed,
       timestamp
FROM gpsStream;
```

Parameter 4326 is the EPSG* identifier of the WGS 84 spatial reference system, and :: is shorthand for typecasting.

---

* http://spatialreference.org/ref/epsg/4326/.

The following is a derived sequence of `mobileObjectWGS84` tuples:

```
...
"W-45084A" POINT(48.20781333 16.43832500 221) 73.2 2016-10-19 11:50:30
"W-45084A" POINT(48.20795500 16.43853167 221) 79.2 2016-10-19 11:50:31
"W-45084A" POINT(48.20809167 16.43873667 220) 77.4 2016-10-19 11:50:32
"W-45084A" POINT(48.20823500 16.43894667 220) 80.3 2016-10-19 11:50:33
"W-45084A" POINT(48.20838167 16.43916333 220) 82.5 2016-10-19 11:50:34
"W-45084A" POINT(48.20851667 16.43936000 220) 75.4 2016-10-19 11:50:35
"W-45084A" POINT(48.20865833 16.43957167 220) 80.1 2016-10-19 11:50:36
"W-45084A" POINT(48.20881000 16.43978667 219) 83.6 2016-10-19 11:50:37
"W-45084A" POINT(48.20894667 16.43997667 219) 74.7 2016-10-19 11:50:38
"W-45084A" POINT(48.20908667 16.44017833 219) 77.8 2016-10-19 11:50:39
"W-45084A" POINT(48.20923167 16.44038333 218) 79.8 2016-10-19 11:50:40
"W-45084A" POINT(48.20937667 16.44059167 218) 80.5 2016-10-19 11:50:41
...
```

We may define another derived data stream with a position modeled as a point in two-dimensional Euclidean space as follows:

```
CREATE STREAM mobileObject AS
SELECT id,
       Force_2D(Transform(wgsPosition::GEOMETRY,3416)) AS position,
       speed,
       timestamp
FROM mobileObjectWGS84;
```

The function `Transform` transforms a point on the WGS84 ellipsoid (`wgsPoint`) into the specified spatial reference system. Parameter 3416 is a unique identifier (SRID) used to unambiguously identify the EPSG:3416[*] spatial reference system, which incorporates European Terrestrial Reference System 1989 (ETRS89) and Lambert projection.

Finally, the following is a tuple sequence of a derived `mobileObject` stream:

```
...
"W-45084A" POINT(630656.02 483284.08) 73.2 2016-10-19 11:50:30
"W-45084A" POINT(630670.74 483300.43) 79.2 2016-10-19 11:50:31
"W-45084A" POINT(630685.35 483316.22) 77.4 2016-10-19 11:50:32
"W-45084A" POINT(630700.30 483332.76) 80.3 2016-10-19 11:50:33
"W-45084A" POINT(630715.74 483349.70) 82.5 2016-10-19 11:50:34
"W-45084A" POINT(630729.74 483365.28) 75.4 2016-10-19 11:50:35
"W-45084A" POINT(630744.82 483381.64) 80.1 2016-10-19 11:50:36
"W-45084A" POINT(630760.17 483399.13) 83.6 2016-10-19 11:50:37
"W-45084A" POINT(630773.62 483414.87) 74.7 2016-10-19 11:50:38
"W-45084A" POINT(630787.97 483431.02) 77.8 2016-10-19 11:50:39
"W-45084A" POINT(630802.54 483447.74) 79.8 2016-10-19 11:50:40
"W-45084A" POINT(630817.36 483464.46) 80.5 2016-10-19 11:50:41
...
```

---

[*] http://spatialreference.org/ref/epsg/3416/.

The previous two examples illustrate the concept of derived streams. Of course, it would be possible to define corresponding geospatial raw streams straightly as follows:

```
CREATE STREAM mobileObjectWGS84 (
      id VARCHAR(8),
      wgsPosition GEOGRAPHY(POINT,4326)
      speed REAL,
      timestamp TIMESTAMP VALIDTIME
)
ORDERED BY timestamp;
ALTER STREAM mobileObjectWGS84 ADD WRAPPER moWrapperWGS84;
```

and

```
CREATE STREAM mobileObject (
      id VARCHAR(8),
      position GEOMETRY(POINT, 3416)
      speed REAL,
      timestamp TIMESTAMP VALIDTIME
)
ORDERED BY timestamp;
ALTER STREAM mobileObject ADD WRAPPER moWrapper;
```

In both cases, the complete logic of data acquisition and transformation of bytes into row streams are encapsulated into the corresponding wrappers.

10.4 Stream Windows

Most DSMSs extend and modify a database query language (such as SQL) to support efficient continuous queries on data streams. As stated before, queries in DSMS run continuously and incrementally produce new results over time. The operators in continuous queries (selection, projection, join, and aggregation, among others) compute on tuples as they arrive and do not presume the data stream is finite, which has significant negative implications. Some operators (Cartesian product, join, union, set difference, and spatial aggregation, among others) require the entire input sets to be completed. These *blocking* operators will produce no results until the data streams end (if ever), which is obviously a serious obstacle and limitation. To get results continuously and not wait until the data streams end, blocking operators must be transformed into *non-blocking* operators. Queries expressible by non-blocking operators are monotonic queries.

Definition 10.9

(*Monotonic query*) A continuous operator or continuous query Q is monotonic if

$$Q(\tau) \subseteq Q(\tau') \quad \text{for all } \tau \leq \tau'$$

Simple selection over a single stream is an example of a monotonic query: at any point in time τ', when a new tuple arrives, it either satisfies selection predicate or it does not, and all the previously returned results (tuples) remain in $Q(\tau')$.

Definition 10.10

(*Non-blocking query*) A non-blocking continuous operator or continuous query $Q^{\neg\ominus}$ is one that produces results (all the tuples of output) before it has detected the end of the input.

Both standard and geospatial aggregate operators always return a stream of length one; they are non-monotonic and thus blocking. Data stream researchers have long recognized the problem of transforming blocking queries into their non-blocking counterpart. A dominant technique that overcomes this problem is to restrict the operator range to a finite window over input streams. Windows were introduced into standard SQL as part of SQL:1999 OLAP functions. In SQL:1999, window is a user-specified selection of rows within a query that determines the set of rows with respect to the current row under examination. The motivation for having the window concept in DSMSs is quite different. Windows limit and focus the scope of an operator or a query to a manageable portion of the data stream. A window is a *stream-to-relation* operator; it specifies a snapshot of a finite portion of a stream at any time point as *a temporary relation*. In other words, windows transform blocking operators and queries to compute in a non-blocking manner. The most recent data are emphasized, which are more relevant than the older data in the majority of data stream applications. There are several window types, though the following two basic types are being extensively used in conventional DSMS: logical, *time-based* windows and physical, *tuple-based* windows. By default, a time-based window is refreshed at every time tick and a tuple-based window is refreshed when a new tuple arrives. The tuples enter and expire from the window in a *first-in-first-expire* pattern; whenever a tuple becomes old enough, it is expired (i.e., deleted) from memory, leaving its space to a more recent tuple. As a result, traditional window queries can support only (recent) historical queries, making them not suitable for geospatial queries concerned with the current state of data rather than recent history. However, these two window types are not useful in answering an interesting and important class of queries over geospatial data streams; therefore, the *predicate-based* window has been proposed in which an arbitrary *logical predicate* specifies the window content.

10.4.1 Time-Based Windows

The time-based window W_ω^T is defined in terms of the window size ω represented as a time interval T_ω. Formally, it takes the stream S and the time interval T_ω as parameters and returns a finite, bounded stream, i.e., a temporary finite relation:

$$W_\omega^T : S \times T_\omega \rightarrow \mathcal{R}$$

The scope of a time-based window W_ω^T denotes the most recent time interval, that is, it consists of the tuples whose timestamp is between the current time τ and $\tau - \omega$.

Let τ_0 denote the time instant that a continuous query specifying a sliding window has effectively started, and τ denote the time instant of the current time. The scope of time interval T_ω may be formally specified as follows:

$$T_\omega(\tau) = \begin{cases} [\tau_0, \tau], & \tau_0 \leq \tau < \tau_0 + \omega \\ [\tau - \omega + 1, \tau], & \tau \geq \tau_0 + \omega \end{cases}$$

The qualifying tuples are included in the window based on their timestamps by appending new tuples and discarding older ones. It is worth noting that a time-based window is refreshed at every time instant, that is, with constant refresh time granularity. We will use the following basic syntax to specify a time-based window W_ω^T on stream S:

`S [RANGE ω]`

EXAMPLE 10.1

A time-based window `gpsStream [RANGE 20 seconds]` defines a window over input stream `gpsStream` with the size of 20 seconds. At any time instant, the window output (relation) contains the bag of tuples from the previous 20 seconds.

There are two important subclasses of time-based windows, *now* window W_{now}^T and *unbounded* window W_∞^T. Now window W_{now}^T, defined by setting $\tau = \text{NOW}$ and $\omega = 1$, returns tuples of stream (relation) with a timestamp equal to NOW. Unbounded window W_∞^T, defined by setting $\omega = \infty$, consists of tuples obtained from all tuples from stream S up to time instant τ. These two special windows are specified using the following syntax:

`S [NOW]`
`S [RANGE UNBOUNDED]`

EXAMPLE 10.2

Suppose that Austrian rivers are stored in a spatial database in a `river` table created as follows:

```
CREATE TABLE river (
     name : VARCHAR(16),
     geometry: GEOMETRY(POLYGON,3416)
)
```

The next query returns a position (type of `Point`) of mobile objects that cross the Danube River at each time instant of the current time:

```
WITH danube AS (
     SELECT geometry FROM river WHERE name = 'Danube'
)
SELECT STREAM position
FROM mobileObject [NOW]
WHERE Crosses(mobileObject.position, danube.geometry)
```

EXAMPLE 10.3

Consider the following query:

```
WITH danube AS (
     SELECT geometry FROM river WHERE name = 'Danube'
)
SELECT STREAM position
FROM mobileObject [RANGE UNBOUNDED]
WHERE Crosses(mobileObject.position, danube.geometry)
```

This query is monotonic and produces a relation that at time τ contains the position of all mobile objects that have crossed the Danube River up to τ.

Time-based windows can optionally contain a *slide* parameter λ, indicating the granularity at which a window slides, that is, how frequently the window should be refreshed. Accordingly, we define the scope of time interval T_ω as follows:

$$T_\omega(\tau) = \begin{cases} [\tau_0, \tau], & \tau_0 \leq \tau < \tau + \omega \wedge & (\tau - \tau_0) \bmod \lambda = 0 \\ [\tau - \omega + 1, \tau], & \tau \geq \tau_0 + \omega \wedge & (\tau - \tau_0) \bmod \lambda = 0 \\ T_\omega(\tau - 1), & & (\tau - \tau_0) \bmod \lambda \neq 0 \end{cases}$$

and use the following syntax construction for sliding a time-based window $W_{\omega,\lambda}^T$:

```
S [RANGE ω SLIDE λ]
```

EXAMPLE 10.4

The following query returns the position of mobile objects that have crossed the Danube River every minute in the past 5 minutes:

```
WITH danube AS (
     SELECT geometry FROM river WHERE name = 'Danube'
)
SELECT STREAM position
FROM mobileObject [RANGE '5 minutes' SLIDE '1 minute']
WHERE Crosses(mobileObject.position, danube.geometry)
```

10.4.2 Tuple-Based Windows

The tuple-based window W^{tuple} defines its output stream over time by a window of the last N elements over its input stream. Formally, it takes a stream S and the natural number $N \in \mathbb{N}^*$ as parameters and returns the following temporary finite relation:

$$W_N^{tuple} : S \times \mathbb{N}^* \to \mathcal{R}$$

At any time instant τ, the output relation $\mathcal{R}$ consists of the N tuples of S with the largest timestamps $\leq \tau$ (or all tuples if the length of up to τ is $\leq N$). We will use the following basic syntax to specify a tuple-based window W^{tuple}:

```
S [ROWS S]
```

The special case of $N = \infty$ is specified by

```
S [ROWS UNBOUNDED]
```

and is equivalent to a time-based window

```
S [RANGE UNBOUNDED]
```

EXAMPLE 10.5

A tuple-based window `gpsStream [ROWS 1]` denotes the "latest" tuple in our `gpsStream`, which is very simple compared to reality. In reality, we will have a number of mobile objects with the same timestamp, and the result of `gpsStream [ROWS 1]` will be ambiguous. As a result, the usability of a tuple-based window in geospatial applications is limited.

10.4.3 Predicate-Based Windows

An important issue in data stream query languages is the frequency with which the answer gets refreshed as well as the conditions that trigger the refresh. Coarser periodic refresh requirements are typically expressed as windows, but users in geospatial applications may not be interested only in refreshing the query answer (i.e., window) in response to every tuple arrival. Consequently, a data stream query language should allow a user to express more general refresh conditions based on arbitrary conditions, including temporal, spatial, and event conditions, among others. For example, consider the following query:

Q^π: Continuously report the position of mobile objects that cross the Danube River.

At any time τ, the window of interest for query Q^π includes only the mobile objects that qualify the predicate *"cross the Danube River."* If a mobile object O reports a position that crosses (i.e., is within) the Danube River, then it should be in Q^π's window. Whenever O reports a position that disqualifies the predicate *"cross the Danube River",* O expires from Q^π's window. It is important to note that objects enter and expire from Q^π's window in an *out-of-order* pattern. An object is expired (and hence is deleted) from Q^π's window only when the object reports another position that disqualifies the window predicate.

The semantics of a *time-based* window query model reads as follows:

Q^T: Continuously report the position of mobile objects that cross the Danube River in the last ω time units.

The query Q^T is semantically different from query Q^π: the window of interest in Q^π includes objects that are *"crossing the Danube River"* while the window of interest in Q^T includes objects that *"have crossed the Danube River in the last ω time units".*

Definition 10.11

(Predicate-based window query) A predicate-based window query Q^π is defined over the data stream S and window predicate Π over the tuples in S. At any point in time τ, the answer to Q^π equals the answer to snapshot query Q_τ^σ, where Q_τ^σ is issued at time τ and the inputs to Q_τ^σ are the tuples in stream S that qualify the predicate Π at time τ.

We will use the following basic syntax to specify the predicate-based window W^Π on stream S:

$$S[\Pi]$$

where Π is the predicate that qualifies and disqualifies tuples into (and out of) the window.

It is important to note that time-based and tuple-based window queries fail to answer some of the predicate-based window queries. Let Q_∞^T denote the query in Example 10.3, which is the *de facto* implementation of query Q^T with $\omega = \infty$. The main difference between a predicate-based window query Q^π

```
WITH danube AS (
      SELECT geometry FROM river WHERE name = 'Danube'
)
SELECT STREAM position
FROM mobileObject AS mo [Crosses(mo.position, danube.
   geometry)]
```

and the query Q_∞^T in Example 10.3 is that a disqualified tuple in the predicate-based window may result in a negative tuple[*] as an output while a disqualified tuple in the WHERE clause predicate of Q_∞^T does not result in any output tuples. When a tuple t qualifies the WHERE predicate of Q_∞^T and is reported in the answer, t will remain in the Q_∞^T query answer forever. In contrast, in the predicate-based window query model, when a tuple t qualifies the predicate Π and is reported in the Q^π answer later, t may be deleted from the query answer if t receives an update so that t does not qualify the window predicate any longer.

Similarly, the *now* window is semantically different from the predicate-based window. Thus, the semantics for the query with the *now* window in Example 10.2 read as follows:

Q^{now}: Report the positions of mobile objects that cross the Danube River now.

At any time, the answer of continuously running query Q^{now} will include only the position of mobile objects that cross the Danube River at time instant $\tau = NOW$. On the other hand, at any time instant τ, the predicate-based window query Q^π may include positions of mobile objects that have crossed the Danube River before τ.

10.5 Distributed Processing of Big Geospatial Data Streams

The growth of data volumes and availability poses tremendous computational challenges. Since data sizes have outpaced the capabilities of centralized architectures and single machines, new applications need new systems to scale computations to distributed architectures and multiple nodes.

Big Data is a term used to identify the datasets that, due to their large size, cannot be managed without using new technologies and programming

---

[*] In the pipelined query execution model with the negative tuples approach, a negative tuple is interpreted as a deletion of a previously produced positive tuple.

frameworks. It does not refer to any specific quantity but rather to high-volume, high-velocity, high-variety, and high-veracity data that demand cost-effective, innovative forms of data processing.

The distributed computing field has achieved success in scaling up big data processing on large numbers of unreliable commodity machines. More recently, the Big Data paradigm has resulted in cluster computing frameworks for large-scale processing, including MapReduce, Hadoop, Hive, and Apache Drill, among others. However, IFP applications do not readily fit their processing models due to an *outbound* (*pull-based*) processing paradigm; dynamic, transient ad hoc queries are typically specified, optimized, and processed once over static, persistent data.

Real-time processing of big geospatial data streams has an essentially different set of requirements than batch processing, and requires a complementary *inbound* (*push-based*) paradigm; the same static, persistent queries are processed continuously over transient, dynamic, frequently changing streaming data.

Early streaming systems are focused on relational style operators for computations, whereas current systems support more general user-defined computations. Stream processing systems commonly define computation over streams as workflow; in a directed acyclic graph (DAG), nodes represent streaming operators in the form of user-defined functions (UDFs) and edges represent execution ordering. Apache Apex, Apache Storm, Apache Spark, and Apache Flink are distributed stream computing platforms capable of processing big data streams but without native support for processing big geospatial data streams.

Geospatial data streams, that is, real-time, transient, time-varying sequences of spatiotemporal data items, demonstrate at least two Big Data core features, *volume* and *velocity*. To handle the volumes of data and computation they involve, these applications need to be distributed over clusters. There is a clear need for highly scalable geospatial stream computing frameworks that can operate at high data rates and process massive amounts of big geospatial data streams. Our goal is to achieve an integrated query processing approach that runs SQL-like expressions continuously and incrementally. The key concept here is that mobility streaming data and persistent data are not intrinsically different; the persistent data are simply streaming data that have been entered into the persistent structures. In other words, query processing could be exclusively performed on persistent data, exclusively on streams, or on a combination of streams and persistent data. In this chapter, we present a framework for distributed big geospatial data stream processing. The framework is a cornerstone toward the efficient real-time management and monitoring of mobile objects through distributed geospatial streams processing on large clusters. A prototype[*] is built on top of a distributed streaming dataflow engine and extends it with a set of spatiotemporal data types and corresponding operations (Figure 10.5). The prototype

---

[*] Available on https://github.com/nkatanic/stFlink.

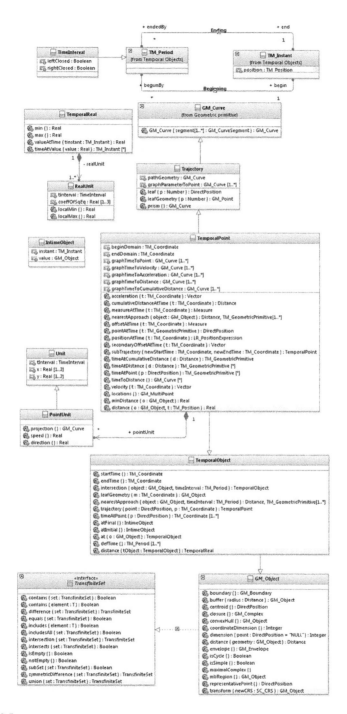

FIGURE 10.5
Data types model—UML class diagram.

implementation is rooted in a new stream-processing model that overcomes the challenges of current distributed stream processing models and enables seamless integration with batch and interactive processing.

10.5.1 Data Model

A data model provides a formalism consisting of a notation for describing data of interest, a set of operations for manipulating these data, and a set of predicates. A spatiotemporal data model is a data model representing the temporal evolution of geospatial objects over time. If these evolutions are continuous, they are mobile objects and represented by spatiotemporal data types such as mobile point (e.g., recording the route of a car), mobile points (e.g., representing movements of a fleet of trucks), and mobile line (e.g., representing movement of a train), among others. A mobile object is a spatiotemporal object that continuously changes its location and/or shape. Depending on the particular mobile objects and applications, the movements of mobile objects may be subject to constraints. In what follows, we focus on *unconstrained*, free movement in Euclidean space.

Our data model (Figure 10.5) is leveraged on ISO (2008) (Güting, Böhlen et al. 2000) and abstract geospatial and temporal data type frameworks.[*]

All spatiotemporal objects share a set of common operations. For this reason, all spatiotemporal data types should implement common operations and operations that are specific for a particular spatiotemporal data type. Similar behavior is expressed through the interface[†] `TemporalObject` that specifies common operations for all spatiotemporal data types. A number of operations, specified by `TemporalObject`, are already defined in the literature; they have the same signature and semantics in our type system as well. For this interface and temporal types `TemporalPoint` and `TemporalPolygon`, we will briefly describe just those operations with different signatures or meaning and the new operations that we propose.

The operation

```
nearestApproach(o:GM_Object, tinterval:TimeInterval):
   TM_Instant[1..*]
```

shall return an array of time instants that lists, in ascending order, the time or times of the nearest approach of the mobile object to a given static geospatial

---

[*] ISO 19108:2002 Geographic information—Temporal schema and ISO 19107:2003: Geographic information—Spatial schema.
[†] Scala programming language does not support interfaces but enables multiple inheritances by implementing interfaces as traits. For this reason, `TemporalObject` is implemented as a trait that enables `TemporalPoint` to inherit from two types.

object. The parameter `tinterval` shall restrict the search to a particular time interval.

The two operations

```
startTime(): TM_Instant
endTime(): TM_Instant
```

return the result of type `TM _ Instant`, which represents the start and end time value of the period in which the type `TemporalPoint` is defined.

The operations `atInitial` and `atFinal` return the type `IntimeObject` with the following two attributes: the position and shape of mobile object in time and space, which are defined with `startTime` operation for `atInitial` and `endTime` operation for `atFinal`.

The `TemporalPoint` is composed of the `PointUnit` set, whereas operations `timeAtPoint`, `subTrajectory`, and `velocity` have the same meaning and signature as the corresponding operations in ISO 19141:2008. The operation `locations` returns isolated points at the beginning and end of each unit's time interval. We introduce operation `distance` exclusively as a measure of linear space between two geospatial objects and operation `length` as a measure of a trajectory's total length. For this reason, we introduce the following operations on `TemporalPoint`:

```
lengthAtTime(tinstant:TM_Instant):Real
timeAtLength(length:Real):TM_Instant
maxDistance(o:GM_Object):Real
minDistance(o:GM_Object):Real
distance(o:GM_Object, tinstant:TM_Instant):Real
```

The operation `lengthAtTime` returns the total trajectory's length from the beginning of the trajectory until the given time instant `tinstant`, and `timeAtLength` returns the time at which the trajectory's total length from the beginning reaches the given value. Operations `minDistance` and `maxDistance` return the distance of the nearest and furthest approach of the temporal point to a given geospatial object. The operation `distance` returns the distance of the temporal point from a given geospatial object at a given time instant `tinstant`.

10.5.2 Apache Flink

As we have already noted, none of the BigData streaming frameworks and platforms have either built-in geospatial or spatiotemporal capabilities. However, we choose Apache Flink as an underlying platform due to its efficient, parallel fault recovery mechanism, and a unique ability to combine nearly real-time and batch in-memory processing in a unified programming framework, which is common in IFP applications involving both real-time and historical mobility data. Another important reason is the fact that we have already noted: geospatial streams are *explicitly* timestamped. As an important consequence, analyzing spatiotemporal events with respect to their *event* time is by far the most interesting compared to analyzing them with respect to the time when they arrive in the system. Therefore, the underlying distributed streaming engine should be configurable for processing geospatial data streams based on event time. The reason for this is that event time for a given tuple is immutable, but processing time changes constantly for each tuple as it flows through the pipeline and time advances.

The core of Apache Flink is a distributed streaming dataflow engine that provides data distribution, communication, and fault tolerance for distributed computations over data streams and executes programs in a data-parallel and pipelined manner. Programs are automatically compiled and optimized into dataflow programs that are executed in a computing cluster or cloud environment. It has advanced the state of the art in data management in several aspects, which are as follows:

1. A data programming model based on second-order functions to abstract parallelization
2. A method that uses static code analysis of user-defined functions to achieve goals similar to database query optimization in a UDF-heavy environment

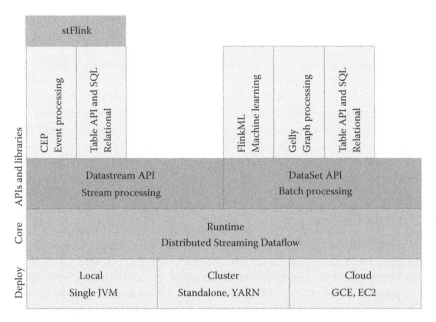

FIGURE 10.6
The Apache Flink stack with Flink library.

3. Abstractions to integrate iterative processing in a dataflow system with good performance, and

4. An extensible query language and an underlying operator model

Apache Flink is a layered system (Figure 10.6); the different layers of the stack build on top of each other and raise the abstraction level of the program representations they accept. DataStream is the basic data abstraction and represents a continuous, parallel, immutable stream of data of a certain type. It is a statically typed object parametrized by an element type, which is itself of the native *Java/Scala* data type.

Event time decouples the program semantics from the actual serving speed of the source and the processing performance of the system. Event time windows compute correct results even if events arrive out of the order of their timestamp, which is common if a data stream gathers events from distributed sources.

10.5.3 Declarative Spatiotemporal Queries

Developing, optimizing, and maintaining complex applications coded in a general-purpose programming language is a hard and resource-consuming task. Major reasons for the widespread success of database systems include *data independence*, separating physical representation and storage from the actual information, and *declarative languages*, separating the program specification from its intended execution environment. In the era of Big Data, many-core processors, distributed computing, and NoSQL ensure that well-established declarative language concepts inherent in relational and object-relational DBMS) that make their way into advanced analytics of big geospatial data streams are of great importance (Markl 2014).

Specification of a declarative spatiotemporal query in our framework is described in Algorithm 10.1.

Algorithm 10.1

Declarative spatiotemporal query involving a time window

Input: $\mathcal{S}_\gamma$,	▷geospatial stream
$\mathcal{W}$	▷time-based window
$k\mathcal{S}_\gamma \leftarrow _{Kkey}\mathcal{S}_\gamma$	▷logically partitioned *keyed* stream
$_{\alpha temporal}w\mathcal{S}_\gamma \leftarrow _{w}k\mathcal{S}_\gamma$	▷windowed stream
$R \leftarrow (R)_{\alpha temporal}w\mathcal{S}_\gamma$	▷convert stream to relation
$R \leftarrow _{\pi f:X \rightarrow Y}R$	▷projection—select clause
$R \leftarrow _{\sigma_{f:B^k \rightarrow B}}R$	▷selection—where clause
return R	

For validation purposes, we used GeoLife[*]: a trajectory of this dataset is represented by a sequence of timestamped points, each of which contains the information of latitude, longitude, and altitude.

We have to specify a data source of a stream with geospatial data stream schema Σ_γ and temporal ordering f_Ω according to *event* time, as we defined before. Therefore, we define a *Scala* case class according to the geospatial data stream schema Σ_γ and use it in a stream execution environment for adding a data source or simply for mapping data read from a socket to a derived data stream:

```
case class sttuple(id:Int, location:Point, eventTime:Timestamp)
```

In this section, we perform an evaluation and validation of our approach by formulating a number of queries that had been specified in natural language before.

Geospatial `DataStream` can be created either from data sources (file-based, collection-based, Apache Kafka, RabbitMQ, Twitter Streaming, etc.) or by applying high-level operations on other `DataStream`.

To work with event time semantics, it is necessary to set execution stream environment time characteristic to an event time:

```
val env = StreamExecutionEnvironment.getExecutionEnvironment
    env.setStreamTimeCharacteristic(TimeCharacteristic.EventTime)
```

To convert the data stream to the relational `Table` abstraction, it is necessary to create an instance of `StreamTableEnvironment`:

```
val tEnv : StreamTableEnvironment = TableEnvironment.getTable
    Environment(env)
```

`DataStream` is directly converted into a `Table` without registering in the `TableEnvironment`, as shown in our query examples.

We also define how timestamps relate to events (e.g., which tuple field is the event timestamp). According to temporal ordering f_Ω of geospatial data streams on event time, we apply the `assignAscendingTimestamps` function:

```
val ststream : DataStream[sttuple] =
    rawstream
    .map{tuple => sttuple(tuple)}
    .assignAscendingTimestamps(tuple => tuple.timeStamp.getTime)
```

---

[*] This GPS dataset was collected by Microsoft Research Asia, that is, by 182 users in a period of over 5 years, from April 2007 to August 2012. It contains approximately 20 million points with a total distance of about 1.2 million kilometers and a total duration of 48,000+h. The data were logged in over 30 cities in China, USA, and Europe.

```
object stFlink {
    def tPoint(stream: DataStream[sttuple],
                window: SlidingWindow
               ): DataStream[temporaltuple] =
        {
          stream
          .keyBy(0)
          .timeWindow(window.size, window.slide)
          .apply {temporal.temporalPoint _}
        }
}
```

Q_1: Continuously report spatiotemporal objects (id and location) within the area of interest:

```
val coords = : Array[Coordinate] = Array (...)
val geomFactory = new GeometryFactory ()
val ring = geomFactory.createLinearRing(coords)
val holes : Array[LinearRing] = null
val areaOfInterest = geomFactory.createPolygon(ring, holes)
```

```
val q1 = ststream
            .toTable(tEnv, 'id, 'point, 'timestamp)
            .select('id, 'point,'timestamp)
            .where(within('point, areaOfInterest()))
```

Q_2: Continuously, each minute, report location of spatiotemporal objects that have traveled more than 3 km in the past 10 minutes:

```
val q2 = stFlink
            .tPoint(ststream, SlidingWindow(Time.minutes(10),
                                            Time.minutes(1)
                                            )
                   )
            .toTable(tEnv, 'id, 'tempPoint)
            .select('driverId, 'tempPoint)
            .where(lengthAtTime('tempPoint,
                                endTime('tempPoint)) > 3000)
```

Q_3: For each spatiotemporal object, find its minimal distance from the point of interest during the past half hour:

```
val easting =…
val northing =…
val point = new Coordinate(easting, northing)
val factory = new GeometryFactory ()
val pointOfInterest = factory.createPoint(point)
```

```
val q3 = stFlink
        .tPoint(ststream, TumblingWindow(Time.minutes(30)))
        .toTable(tEnv, 'id, 'tempPoint)
        .select('id,
                minDistance('tempPoint, pointOfInterest())
                as 'minimalDistance
                )
```

Q_4: Find all spatiotemporal objects (id and distance traveled) that have traveled more than 10 km during the past hour.

```
val q4 = stFlink
        .tPoint(ststream, TumblingWindow(Time.minutes(60)))
        .toTable(tEnv, 'id, 'tempPoint)
        .select('id,
                'tempPoint,
                lengthAtTime('tempPoint, endTime('tempPoint)
                        ) as 'distanceTraveled
                )
        .where('distanceTraveled > 10000 )
```

Q_5: Find trajectories of the spatiotemporal objects that have been less than 500 m from a point of interest within the past 5 minutes.

```
val q5 = stFlink
        .tPoint(ststream, TumblingWindow(Time.minutes(5)))
        .toTable(tEnv, 'id, 'tempPoint)
        .select('id,
                'tempPoint,
                subTrajectory('tempPoint,
                                startTime('tempPoint),
                                endTime('tempPoint)
                        ) as 'subtrajectory
                )
        .where(distance('tempPoint, pointOfInterest(), endTime
        ('tempPoint)
                        ) < 500
                )
```

10.6 Summary

Geospatial stream processing refers to a class of software systems for processing high-volume geospatial data streams with very low latency, that is, in near real time. Motivated by the limitation of DBMS, the database community developed DSMSs as a new class of management systems oriented toward processing large data streams. Despite differences between these two classes of management systems, DSMSs resemble DBMSs; they process data streams using SQL and operators defined by relational algebra.

Geospatial data streams, that is, real-time, transient, time-varying sequences of spatiotemporal data items, demonstrate at least two Big Data core features, *volume* and *velocity*. To handle the huge volumes of data streams and computations they involve, these applications need to be distributed over clusters. However, despite substantial work on cluster programming models for batch computation, there are few similar high-level tools for geospatial stream processing. There is a clear need for a highly scalable stream computing framework that can operate at high data rates and process massive amounts of geospatial data streams. In this chapter, we presented an insight into geospatial stream processing at a conceptual level as well as an approach and framework for geospatial data streams processing using a distributed stream processing engine.

Bibliographic Notes

The literature on data streams is rather extensive. The introductory Section 2 borrows material from Babcock et al. (2002), Golab and Özsu (2010), and Stonebraker, Çetintemel, and Zdonik (2005). IFP and related terms (IFP application domain, IFP engine, and IFP systems) were introduced in Cugola and Margara (2012).

The definition of temporal order and time-based sliding windows can be found in the paper by Patroumpas and Sellis (2011), whereas the definition of monotonic query and non-blocking query can be found in Golab and Özsu (2010) and Law, Wang, and Zaniolo (2011). Readers interested in a precise definition of window semantics in continuous queries are referred to Patroumpas and Sellis (2006). Predicate windows were introduced in Ghanem, Aref, and Elmagarmid (2006) and extended in Ghanem, Elmagarmid et al. (2010).

Several references exist for continuous query languages, including Arasu, Babu, and Widom (2006), Jain et al. (2008), and Law, Wang, and Zaniolo (2011).

Background knowledge on geostreaming and spatiotemporal data streams can be found in Mokbel and Aref (2008), Huang and Zhang (2008), Kazemitabar, Banaei-Kashani, and McLeod (2011), and Galić, Baranović et al. (2014). Most of the geospatial data stream processing issues from the user perspective were addressed in (Galić 2016), which nicely complements the content of this chapter.

A data model and query languages for mobile objects are covered in depth in the textbook by Gütting and Schneider (2005). The data model of Section 10.4 extends the work presented in (Güting, Böhlen et al. 2000) and (ISO 2008).

Although this chapter does not focus on optimization issues, interested readers are referred to Elmongui, Ouzzani, and Aref (2006), who present several major challenges related to the lack of spatiotemporal pipelined operators and the impact of time, space, and their combination on query optimization. The works presented in Mahmood et al. (2015) and Abdelhamid et al. (2016) focus on extending distributed stream engines with the adaptive indexing layer and adaptive main-memory data partitioning query processing technique.

Bibliography

Abdelhamid, A. S., M. Tang, A. M. Aly, A. R. Mahmood, T. Qadah, W. G. Aref, and S. Basalamah 2016. Cruncher: Distributed in-memory processing for location-based services. *32nd IEEE International Conference on Data Engineering, ICDE 2016*. New York: IEEE Computer Society, 1406–1409.

Arasu, A., S. Babu, and J. Widom. 2006. The CQL continuous query language: Semantic foundations and query execution. *VLDB Journal* 15 (2): 121–142.

Babcock, B., S. Babu, M. Datar, R. Motwani, and J. Widom. 2002. Models and issues in data stream systems. *Proceedings of the 21st ACM SIGACT-SIGMOD-SIGART Symposium on Principles of Database Systems*. New York: ACM, 1–16.

Cugola, G. and A. Margara. 2012. Processing flows of information: From data stream to complex event processing. *ACM Computing Surveys* 44 (3): 15:1–15:60.

Elmongui, H. G., M. Ouzzani, and W. G. Aref. 2006. *Challenges in spatiotemporal stream query optimization. Fifth ACM International Workshop on Data Engineering for Wireless and Mobile Access*. New York: ACM. 27–34.

Güting, R. H. and M. Schneider. 2005. *Moving Objects Databases*. San Francisco, CA: Morgan Kaufman Publishers Inc.

Güting, R. H., M. H. Böhlen, M. Erwig, C. S. Jensen, N. A. Lorentzos, M. Schneider, and M. Vazirgiannis. 2000. A foundation for representing and querying moving objects. *ACM Transactions on Database Systems* 25 (1): 1–42.

Galić, Z. 2016. *Spatio-Temporal Data Streams*. New York, NY: Springer.

Galić, Z., M. Baranović, K. Križanović, and E. Mešković. 2014. Geospatial data streams: Formal framework and implementation. *Data & Knowledge Engineering* 91: 1–16.

Ghanem, M. T., W. G. Aref, and A. K. Elmagarmid. 2006. Exploiting predicate-window semantics over data streams. *SIGMOD Record* 35 (1): 3–8.

Ghanem, M. T., A. K. Elmagarmid, P-Å. Larson, and W. G. Aref. 2010. Supporting views in data stream management systems. *ACM Transactions on Database Systems* 35 (1): 1:2–1:47.

Golab, L. and M. T. Özsu. 2010. *Data Stream Management*. San Rafael, CA: Morgan Claypool Publishers.

Huang, Y. and C. Zhang. 2008. New data types and operations to support geo-streams. *Geographic Information Science, 5th International Conference, GIScience 2008*. Berlin: Springer. 106–118.

ISO. 2008. *ISO 19141:2008 Geographic Information—Schema for Moving Features*. Geneva: ISO.

Jain N. et al. 2008. Towards a streaming SQL standard. *Proceedings of the VLDB Endowment* 1 (2): 1379–1390.

Kazemitabar, S. J., F. B. Banaei-Kashani, and D. McLeod. 2011. Geostreaming in cloud. *Proceedings of the 2011 ACM SIGSPATIAL International Workshop on GeoStreaming, IWGS 2011*. New York: ACM, 3–9.

Law, Y.- N., H. Wang, and C. Zaniolo. 2011. Relational languages and data models for continuous queries on sequences and data streams. *ACM Transactions on Database Systems* 36 (2): 8:1–8:32.

Mahmood, A. R. et al. 2015. Tornado: A distributed spatio-textual stream processing system. *Proceedings of the VLDB Endowment* 8 (12): 2020–2023.

Mokbel, F. M. and G. W. Aref. 2008. SOLE: Scalable on-line execution of continuous queries on spatio-temporal data streams. *VLDB Journal* 17 (5): 971–995.

Patroumpas, K. and K. T. Sellis. 2011. Maintaining consistent results of continuous queries under diverse. *Information Systems* 36 (1): 42–61.

Patroumpas, K. and K. T. Sellis. 2006. *Window specification over data streams. Proceedings of the 2006 International Conference on Current Trends in Database Technology—EDBT 2006*. New York: Springer, 445–464.

Stonebraker, M., U. Çetintemel, and S. B. Zdonik. 2005. The 8 requirements of real-time stream processing. *SIGMOD Record* 34 (4): 42–47.

Volkerl, M. 2014. Breaking the chains: On declarative data analysis and data independence in the Big Data era. *Proceedings of VLDB Endowment* 7 (13): 1730–1733.

Index

#0001 - 071017 - C28 - 234/156/15 [17] - CB - 9781138626447